Mathematics and Visualization

Series Editors

Gerald Farin
Hans-Christian Hege
David Hoffman
Christopher R. Johnson
Konrad Polthier
Martin Rumpf

Øyvind Hjelle
Morten Dæhlen

Triangulations and Applications

With 126 Figures

 Springer

Øyvind Hjelle
Simula Research Laboratory AS
P.O. Box 134
1325 Lysaker, Norway
email: oyvindhj@simula.no

Morten Dæhlen
Department of Informatics
University of Oslo
P.O. Box 1080, Blindern
0316 Oslo, Norway
email: mortend@ifi.uio.no

Library of Congress Control Number: 2006928289

Mathematics Subject Classification: 51-02

ISBN-10   3-540-33260-X Springer Berlin Heidelberg New York
ISBN-13   978-3-540-33260-2 Springer Berlin Heidelberg New York

Springer is a part of Springer Science+Business Media
springer.com
© Springer-Verlag Berlin Heidelberg 2006
Printed in The Netherlands

Typesetting: by the authors and techbooks using a Springer LATEX macro package
Cover design: *design & production* GmbH, Heidelberg

Printed on acid-free paper   SPIN: 11693598   46/techbooks   5 4 3 2 1 0

Preface

This book is entirely about triangulations. With emphasis on computational issues, we present the basic theory necessary to construct and manipulate triangulations. In particular, we make a tour through the theory behind the Delaunay triangulation, including algorithms and software issues. We also discuss various data structures used for the representation of triangulations. Throughout the book we relate the theory to selected applications, in particular surface construction, meshing and visualization.

The field of triangulation is part of the huge area of computational geometry, and over many years numerous books and articles have been written on the subject. Important results on triangulations have appeared in theoretical books and articles, mostly within the realm of computational geometry. However, many important results on triangulations have also been presented in publications within other research areas, where they have played and play an important role in solving specific scientific and applied problems. We will touch upon some of these areas in this book.

Triangulations, almost everywhere. The early development of triangulation comes from surveying and the art of constructing maps – cartography. Surveyors and cartographers used triangles as the basic geometric feature for calculating distances between points on the Earth's surface and a position's elevation above sea level.

Since the early development of computers an enormous effort has been put into theory, numerical methods and various algorithms for constructing and handling triangulations. This book gives most of the important results with respect to a family of triangulations that fulfill certain criteria. The most important type of triangulation we are investigating, the Delaunay triangulation, is named after the Russian mathematician Boris N. Delaunay, who first described this important triangulation scheme in 1934. During recent decades, advances in computer hardware and software have also brought triangulation technology into many new areas of application. Together with some new results, we cover the major achievements of Delaunay triangulation methods and some important applications of such triangulations.

Triangle-based surfaces, or surfaces represented over triangulations, are used in a wide range of applications. They are to be found in computer aided design (CAD) systems in the automotive industry, and they are used for defining meshes and geometry needed in systems for simulating processes and phenomena based on finite element methods (FEM). Moreover, they are extensively used in systems for representing the geometry of geological structures, and in medical applications triangulations are used for representing the anatomy of the human body. Triangulations are also used in geographical information systems (GIS), mainly for the purpose of representing parts of the Earth's surface, more commonly referred to as terrain models or triangular irregular networks (TIN). In surveying and cartography, triangulations have been used since the ancient Greeks for measuring the Earth's surface, and most of the early results on the subject were developed by practical cartographers for creating maps. Finally, triangulations have become one of the main features in visualization and computer graphics. In particular, the game industry has pushed the development of graphics hardware in order to obtain real-time visualization of as huge data sets as possible. Triangulations have been of particular interest to the developers of graphics hardware, mainly because of their ultimate simplicity – a triangulation is a collection of triangles, and a triangle is given by three points in space. Three points are the minimum number of points needed to represent a piece of a surface in space.

Measurement data, in many cases also referred to as scattered data, are results of spatial measurements taken from some physical body, e.g. the Earth's surface. Data are collected from satellite images, aeroplane images, the global positioning system (GPS) and other types of equipment for measuring position. Geological data most often comes from seismic surveys and well measurements. In medicine, data can be extracted from ultrasound images and images created by magnetic resonance imaging (MRI). A common challenge in all these areas is to construct the underlying geometry of the measured object, and many systems today use triangles and triangulations for this purpose.

Although the theory presented in this book is general and we are aware of the wide variety of applications which use triangulations, we use surface construction and meshing as the main examples throughout the book. In particular, we are interested in the construction and manipulation of triangulations based on some suitable set of measured data.

Teaching triangulations. This book is based on lecture notes from a course on triangulation given at graduate level at the Department of Informatics at the University of Oslo. The course is given over one semester and tailored to be one third of a full-time student's workload during the semester. The semester typically starts mid August with oral exam sometime during December. The course is given over 12–14 double lectures and a number of programming exercises based on the companion software. Except for the final chapter on software, the organization of the chapters is based on the chronology of our lectures. The students are introduced to the companion software when needed as programming exercises are given throughout the course.

How to read this book. Although our advice is to follow the chronology of the chapters, the reader can also follow other paths through this book.

- An introduction to triangles and triangulations and the basic theory of Delaunay triangulations are given in Chapters 1, 3 and 4. In Chapter 3 we define the Delaunay triangulation and in Chapter 4 we discuss the most important algorithms for constructing the Delaunay triangulation. In Chapter 1 we give a brief introduction to triangles and triangulations and give some necessary basic properties of triangulations.
- Chapters 5, 6 and 7 contain three types of triangulations based on the theory of Delaunay triangulation. All three chapters are heavily based on the basic theory given in Chapters 3 and 4. In Chapter 5, we describe methods for constructing triangulations that depend on the shape or the behavior of the given data. In Chapter 6, we introduce constraints in the triangulation itself by keeping edges of triangles fixed, and in Chapter 7, we construct triangle-based meshes suited for finite element calculations.
- In Chapter 8, we use triangulations as a basis for constructing surfaces from huge data sets. This chapter can be read either separately or after a brief look at Chapter 1.
- Chapters 2 and 9 are directed towards implementation issues. In Chapter 2, we discuss data structures for representing triangulations and Chapter 9 is dedicated to generic software components for constructing and manipulating triangulations.

Another possible path through the book is as follows: start with Chapters 1, 3 and 4, continue with Chapters 2 and 9, then read Chapter 8, before you end the course with a selection from the contents of Chapters 5, 6 and 7.

Related topics. As pointed out above, this book is based on lecture notes for a graduate level course at the Department of Informatics at the University of Oslo. In order to give the students a more visual understanding of triangulations, we usually add some lectures on visualization and graphics, including a short introduction to OpenGL [1, 89].

We have decided not to give a detailed introduction to simplification and refinement of triangular meshes, except for the specific mesh generation scheme discussed in Chapter 7. However, insertion and deletion of points in triangulations are covered by the general theory and can therefore be regarded as a basis for both simplification and refinement. For those who are particularly interested in simplification of triangular meshes, see for example [26].

In [19], de Berg, van Kreveld, Overmars and Schwarzkopf give a thorough introduction to computational geometry where they also cover important aspects of triangulations and algorithms for handling triangular meshes. In [26], H. Edelsbrunner covers topics in geometry and topology applied to grid and mesh generation where the theory behind Delaunay triangulations and Voronoi diagrams is also discussed. In addition to surface simplification as mentioned above, Edelsbrunner also covers tetrahedral meshes, which are natural extensions of triangulations to three dimensions.

Software companion. An important feature of this book is the companion software, the Triangulation Template Library (TTL). TTL is open source software and can be downloaded from `www.simula.no/ogl/ttl`. The main ideas behind TTL and description of its functionality are given in the last chapter of this book. We present a generic programming philosophy for triangulations with a clear separation of algorithms and data structures. Since TTL is frequently extended and changed, we have decided not to exploit TTL in too much detail in this book. Important software issues and implementation aspects can be found on the TTL homepage.

Acknowledgement. As mentioned above, this book is based on lecture notes, and first of all we would like to thank all the students who have attended the course and have given us continuous feedback on the contents, outline and details as the lecture notes have evolved. In particular, we wish to thank students Thomas Elboth, Øystein Aanrud, Stein Grongstad, Tom F. Blenning Klaussen, Siri Øyen Larsen, Per-Idar Evensen and Philip Bruvold for their detailed feedback. In addition, Per-Idar Evensen has done excellent work on testing the software as a part of his Master thesis. A special thank you goes to Thomas Sevaldrud who has provided us with valuable comments on the contents, for the production of computer graphics examples and for extensive use and feedback on the software. We wish to thank Hans Petter Langtangen, Kjell Kjenstad, Kyrre Strøm and Martin Reimers for their comments and valuable feedback on specific parts of the manuscript. Finally, we would like to thank Professor Jonathan Shewchuk at Berkeley for letting us use his brilliant illustrations on Delaunay meshing in Chapter 7. We also thank SINTEF Applied Mathematics for supporting the development of the Triangulation Template Library (TTL). The cover picture is taken from the commercial flight simulator Silent Wings (www.silentwings.no) that uses our companion software TTL for the construction of the triangle-based terrain surface.

Norway   *Øyvind Hjelle*
May 2006   *Morten Dæhlen*

Contents

1

Triangles and Triangulations

Triangulations are widely used as a basis for representing geometries and other information appearing in a huge variety of applications. In geographical information systems (GIS), triangulations are used to represent terrain surfaces and relations between geographical objects. Systems for modeling geological structures in the oil and gas industry use triangulations for representing surfaces that separate different geological structures, and for representing properties of these structures. Computer aided design (CAD) systems with triangulation features are common in the manufacturing industry and in particular within the automotive industry, which has been a driving force for this research for many decades. We also find applications using triangulations within engineering fields that simulate physical phenomena using finite element methods (FEM). Finally, we will mention the importance of triangulation in visualization and computer graphics.

This introductory chapter serves two purposes. First we give an introduction to triangulations through some basic thoughts on triangles and how triangles can be organized to form a triangulation. Secondly, we introduce some notation and definitions concerning triangulations, including a simple algorithm for constructing a triangulation from a set of points in the plane. Some basic properties of triangulations are outlined for use in succeeding chapters.

1.1 Triangles

A triangulation is made up of a collection of triangles, and as the word Suggests, a triangle has three angles. Draw two straight lines on a piece of paper, but make sure that the two lines are not parallel to each other. Somewhere, hopefully not outside your sheet of paper, the lines intersect. Add a third line, not parallel to any of the first two lines and not through the intersection point between the first two lines. Your sheet of paper is now divided into seven

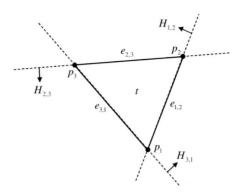

Fig. 1.1. Three points, three edges and three angles – a triangle.

regions, one of which is a geometric object with three corners, three edges and three angles. We have a triangle as illustrated in Figure 1.1.

Another and more common way to construct a triangle is to mark three points on a paper. However, make sure that the three points do not fall on one straight line - the three points must not be collinear. Consider two and two points and draw straight lines between them. You obtain a geometric object with corners at the three points, three edges between two and two points, and three angles. We obtain a triangle. Hence, a triangle is uniquely defined by three non-collinear points p_1, p_2 and p_3, ($p_i = (x_i, y_i)$) in the plane. As a convention we number the points counterclockwise around the triangle. The edge between p_i and p_j we denote by e_{ij} or $e_{i,j}$ as in Figure 1.1. We can also obtain a triangle t as the intersection between three half-planes,

$$t = H_{1,2} \cap H_{2,3} \cap H_{3,1},$$

where $H_{i,j}$ is the half-plane containing t and which is defined by the triangle edge $e_{i,j}$, see Figure 1.1.

When constructing and manipulating triangulations we often need to investigate each triangle by calculating some of its properties, and we conclude this section with a few properties that are useful when constructing triangulations.

The *circumcircle*, or *circumscribed circle* of a triangle is the unique circle through the three points p_1, p_2 and p_3. The triangle is contained within the circular disk generated by this circle. The center of the circle is the unique point c in the plane that is equally distant from p_1, p_2 and p_3. Between p_i and p_j there exists a line $l_{i,j}$, where all points on $l_{i,j}$ have the same distance from p_i and p_j. For a triangle, we can produce three such bisecting lines as illustrated in Figure 1.2, and the three lines intersect in one point c which has the same distance to all three points. This must be the center of the circumcircle, or the *circumcenter*, of the triangle formed by the three points

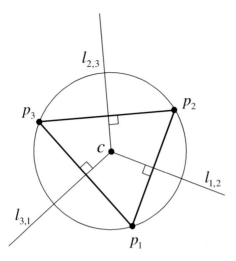

Fig. 1.2. A triangle with circumcircle and bisecting lines.

p_1, p_2 and p_3. We observe that the line $l_{i,j}$ is perpendicular to the edge $e_{i,j}$. (See also Exercise 3.)

Another important issue when constructing triangulations is to calculate the angles of a triangle, and in particular we are interested in the smallest of the three angles. For reasons which will be explained later, we will often try to avoid triangles with small angles. In general, we can apply the laws of sines and cosines. Let a, b and c be the length of the three edges and let α, β and γ be the opposite angles, respectively, see Figure 1.3. We have the following two formulas that give the angles of a given triangle:

$$\frac{\sin \alpha}{a} = \frac{\sin \beta}{b} = \frac{\sin \gamma}{c}$$

$$c^2 = a^2 + b^2 - 2ab \cos \gamma$$

Formulas for calculating sine and cosine from points in the plane are given in Section 3.7.

There are many other interesting properties of triangles. The center of the inscribed circle is found by intersecting the straight lines which bisect each of the angles of a triangle. The area of a triangle with corner points p_1, p_2 and p_3, $(p_i = (x_i, y_i))$ is given by

$$A(p_1, p_2, p_3) = \frac{1}{2} \begin{vmatrix} x_1 & y_1 & 1 \\ x_2 & y_2 & 1 \\ x_3 & y_3 & 1 \end{vmatrix} = \frac{1}{2} (x_1 y_2 + x_2 y_3 + x_3 y_1 - x_1 y_3 - x_2 y_1 - x_3 y_2)$$

where $|\cdot|$ denotes the determinant.

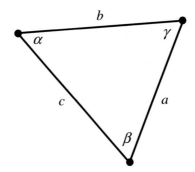

Fig. 1.3. Edges and angles of a triangle.

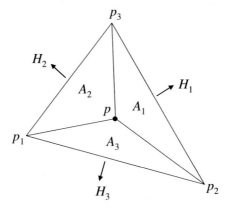

Fig. 1.4. Subdivision of a triangle for calculating barycentric coordinates of a point p.

Any point p in the plane can be written as a linear combination of the three vertices p_1, p_2 and p_3 of a triangle,

$$p = b_1 p_1 + b_2 p_2 + b_3 p_3.$$

If $b_1 + b_2 + b_3 = 1$, then the triple (b_1, b_2, b_3) is known as the barycentric coordinates of p with respect to p_1, p_2 and p_3. Geometrically, the barycentric coordinates can be expressed as ratios of the areas shown in Figure 1.4 and the total triangle area $A = A_1 + A_2 + A_3$,

$$b_1 = \frac{A_1}{A}, \quad b_2 = \frac{A_2}{A} \quad b_3 = \frac{A_3}{A}. \tag{1.1}$$

If $b_1, b_2, b_3 < 1$ then p is strictly inside the triangle. If $b_i = 0$ then p lies on the triangle edge opposite to p_i or on the extension of this edge, and if $b_i = 1$ then p coincides with p_i. Let H_i be the half-plane strictly outside the triangle defined by the edge on the opposite side of p_i as indicated in the figure. If p lies in H_i, then $b_i < 0$. Thus, A_i is a signed area which is less than zero when p lies in H_i.

We also know that the sum of the interior angles of a triangle is equal to π (180°). When necessary, we will come back to these and other properties of triangles. However, our main focus in this book is to investigate triangulations, which are collections of triangles.

1.2 Triangulations

For organizing the collection of triangles in a triangulation and to make triangulations easy to handle, it is necessary to impose certain restrictions. More precisely, we must enforce restrictions so that a triangulation Δ becomes a subdivision of a domain into a collection of connected non-overlapping triangles. The triangles of a triangulation are formed by points given in the domain Ω of interest, see Figure 1.5. These points can either be given or selected by some suitable procedure.

In most cases when constructing triangulations, we start with a given collection of points, say

$$P = \{p_i\}, \ i = 1, ..., N,$$

and a domain Ω, which contains all the points in P. We assume that the boundary of Ω is one or more closed *simple polygons*. A simple polygon is a polygon that does not self-intersect. In many cases we prefer Ω to be the *convex hull* of the point set.

Definition 1.1 (Convex hull and convex set). *The convex hull of a set of points P is the smallest convex set containing P. A set S is convex if any line segment joining two points in S lies entirely in S.*

Figure 1.5 shows a convex domain Ω with a set of points in the plane, and a corresponding triangulation of these points. We denote the boundary of Ω by $\partial\Omega$.

In the following we use the word *point* when referring to the geometric position in the plane of a vertex in a triangulation. The vertex (or node) of a triangulation denotes a topological element, which does not necessarily hold a geometric position. The topology of a triangulation is about relations between its vertices, edges and triangles. We associate the points p_i, p_j, and p_k with the vertices v_i, v_j and v_k. A single triangle $t_{i,j,k}$ (or t_{ijk}) in a triangulation Δ is spanned by three vertices v_i, v_j and v_k. We assume that the triple (i, j, k) is ordered such that the vertices are arranged counterclockwise around the triangle. At this point we assume that the edges are not ordered, that is $e_{i,j}$ and $e_{j,i}$ represent the same edge between v_i and v_j.

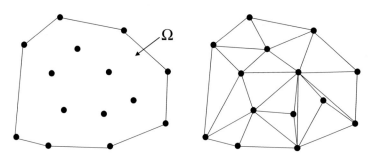

Fig. 1.5. A domain with points and a triangulation.

For representing the entire triangulation, we also introduce the set of triples I_Δ representing all the triangles in the triangulation, that is, the triple

$$\mathbf{i} = (i, j, k) \in I_\Delta$$

for some integers i, j, and k, refers to the triangle $t_{i,j,k}$ in the triangulation Δ.

In general, we could call any collection of triangles in the plane a triangulation. However, for practical and theoretical reasons we are interested in a family of triangulations that meets the following requirements:

1. No triangle $t_{i,j,k}$ in a triangulation Δ is degenerate, that is, if $(i, j, k) \in I_\Delta$, then p_i, p_j, and p_k are not collinear.
2. The interiors of any two triangles in Δ do not intersect, that is, if $(i, j, k) \in I_\Delta$ and $(\alpha, \beta, \gamma) \in I_\Delta$, then

$$Int(t_{i,j,k}) \cap Int(t_{\alpha,\beta,\gamma}) = \phi.$$

3. The boundaries of two triangles can only intersect at a common edge or at a common vertex.
4. The union of all triangles in a triangulation Δ is equal to the domain over which the triangulation is defined, that is

$$\Omega = \cup t_{i,j,k}, \ (i, j, k) \in I_\Delta.$$

5. The domain Ω must be connected.
6. The triangulation shall not have holes.
7. If v_i is a vertex at the boundary $\partial\Omega$, then there must be exactly two boundary edges that have v_i as a common vertex. This implies that the number of boundary vertices is equal to the number of boundary edges.

If the first four requirements are fulfilled, a triangulation is often called a *valid* triangulation. However, throughout this book we will also refer to so-called *regular* triangulations. To ensure regularity, we need to satisfy the three

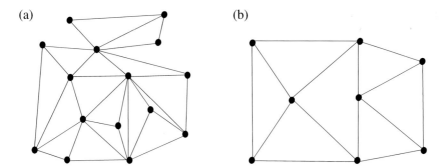

Fig. 1.6. (a) A valid triangulation that is not regular, (b) a triangulation that is not valid.

last requirements in addition to the first four. The triangulation in Figure 1.5 is valid and regular according to the requirements above. Figure 1.6(a) shows a valid triangulation that is not regular since more than two boundary edges meet at one of the boundary vertices. The triangulation in part (b) of the figure is not valid according to requirement (3), hence not regular either.

From a computational point of view it is important to find a suitable representation of a triangulation, hence we need an adequate data structure in our computer programs. Since triangulations are used in many different applications and very often serve different purposes with respect to these applications, various types of data structures are needed. Chapter 2 of this book is devoted to how triangulations can be represented to serve these purposes.

We have explained what we mean by a triangulation and imposed certain restrictions in order to make triangulations more convenient to handle. The next topic is to study some properties important for designing data structures and developing computer programs operating on these structures.

1.3 Some Properties of Triangulations

Triangulations have many interesting properties that can be deduced directly from the properties of planar graphs as described in elementary graph theory [58]. An important issue, in particular with respect to designing data structures and computer programs operating on these structures, is to know "the size of a triangulation". In this section we give some important results on the relation between the number of vertices, edges and triangles. A complete account of these properties is important when implementing algorithms for constructing triangulations and algorithms for traversing a triangulation. We will establish some of these properties for later use.

First we derive important relations between the number of vertices, edges and triangles in a regular triangulation. Let V, E and T be the sets of vertices, edges and triangles in a triangulation. We use subscripts $_I$ and $_B$ to denote interior elements and boundary elements respectively, for example, V_I is the set of interior vertices and V_B is the set of vertices at the boundary. Hence, the total set of vertices in a triangulation is $V = V_I \cup V_B$. The number of elements in the sets V, E and T we denote by $|V|$, $|E|$ and $|T|$, respectively.

Lemma 1.1. *For a regular triangulation Δ we have*

$$|T| = 2|V_I| + |V_B| - 2 \tag{1.2}$$
$$|E| = 3|V_I| + 2|V_B| - 3 \tag{1.3}$$
$$|E_I| = 3|V_I| + |V_B| - 3. \tag{1.4}$$

Proof. We easily observe that the formulas hold when there is only one triangle in Δ; then $|V_I| = 0$ and $|V_B| = 3$. The general cases can be proved by induction. Assume $|T| \geq 2$ and that one triangle t_i is removed from the boundary of Δ in such a way that the reduced triangulation $\widehat{\Delta}$ is also regular. To prove (1.2) there are two cases to be considered.

1. If T_i has two boundary edges in Δ we have

$$|V_I(\widehat{\Delta})| = |V_I(\Delta)|, \text{ and } |V_B(\widehat{\Delta})| = |V_B(\Delta)| - 1.$$

Since we assume that (1.2) holds for $\widehat{\Delta}$ we obtain

$$|T(\Delta)| = |T(\widehat{\Delta})| + 1 = \left(2|V_I(\widehat{\Delta})| + |V_B(\widehat{\Delta})| - 2\right) + 1$$
$$= 2|V_I(\Delta)| + (|V_B(\Delta)| - 1) - 2 + 1$$
$$= 2|V_I(\Delta)| + |V_B| - 2.$$

2. If T_i has one boundary edge in Δ we have

$$|V_I(\widehat{\Delta})| = |V_I(\Delta)| - 1, \text{ and } |V_B(\widehat{\Delta})| = |V_B(\Delta)| + 1.$$

Since we assume that (1.2) holds for $\widehat{\Delta}$ we obtain

$$|T(\Delta)| = |T(\widehat{\Delta})| + 1 = \left(2|V_I(\widehat{\Delta})| + |V_B(\widehat{\Delta})| - 2\right) + 1$$
$$= 2(|V_I(\Delta)| - 1) + (|V_B(\Delta)| + 1) - 2 + 1$$
$$= 2|V_I(\Delta)| + |V_B| - 2.$$

We leave the proofs of (1.3) and (1.4) as an exercise, see Exercise 8.   □

Corollary 1.1. *A special case of the Euler Polyhedron Formula, also known as the Euler-Poincaré formula,*

$$|T| = |E| - |V| + 1, \tag{1.5}$$

follows by combining (1.2) and (1.3).

From the formulas in Lemma 1.1 we observe that once the boundary of the triangulation is specified, the total number of edges and triangles is fixed. Moreover, adding an interior point in an existing triangulation increases the number of triangles and edges by two and three, respectively. As we shall see in later chapters the equations derived above are useful when checking the topological consistency of a triangulation.

From Lemma 1.1, lower and upper bounds on the number of triangles and edges are easily obtained in terms of the number of vertices.

Lemma 1.2. *The number of triangles $|T|$ and edges $|E|$ in a triangulation with $|V|$ vertices satisfies the following relations:*

$$|V| - 2 \leq |T| \leq 2\,|V| - 5 \tag{1.6}$$
$$2\,|V| - 3 \leq |E| \leq 3\,|V| - 6. \tag{1.7}$$

Proof. The proof follows from (1.2) and (1.3) using $|V| = |V_I| + |V_B|$, and the fact that the number of boundary vertices satisfies $|V_B| \geq 3$ and $|V_B| \leq |V|$. $\square$

Another important property relates to the *degree*, also referred to as the *valency*, of a vertex v_i in a triangulation Δ. The degree is denoted by $\deg(v_i)$, and defined as the number of edges *incident with* v_i in Δ. That is, the number of edges joining v_i with another vertex in Δ.

Lemma 1.3. *The sum of the degrees of a triangulation (regular or not) satisfies*

$$\sum_{i=1}^{|V|} \deg(v_i) = 2\,|E|. \tag{1.8}$$

Proof. The proof follows from the observation that every edge is incident with exactly two vertices. $\square$

When constructing a triangulation Δ from a large number of points such that the boundary of Δ is the convex hull of the points, the number of vertices at the boundary will typically be much smaller than the total number of vertices. Thus, combining (1.8) and (1.3) we find that the sum of the degrees of all vertices of a triangulation is approximately six times the number of vertices:

$$\sum_{i=1}^{|V|} \deg(v_i) = 2\,|E| = 6\,|V_I| + 4\,|V_B| - 6 = 6\,|V| - 2\,|V_B| - 6 \approx 6\,|V|, \quad (1.9)$$

when $|V_B|$ is small compared to $|V|$. In other words, the average number of edges incident with a vertex in a triangulation is six, given the conditions above. Under the same conditions we can also obtain estimates of the number of triangles and edges in a triangulation. Using (1.2) and (1.3) we obtain

$$|T| = 2\,|V_I| + |V_B| - 2 = 2\,|V| - |V_B| - 2 \approx 2\,|V| \quad \text{and} \qquad (1.10)$$
$$|E| = 3\,|V_I| + 2\,|V_B| - 3 = 3\,|V| - |V_B| - 3 \approx 3\,|V|\,, \qquad (1.11)$$

if the total number of vertices $|V|$ is much greater than the number of vertices $|V_B|$ at the boundary. These estimates are useful when determining global costs of algorithms for traversing triangulations, and when estimating storage requirements for data structures.

So, a rule of thumb that is useful to have in mind is this: *there are approximately two triangles per vertex and approximately three edges per vertex – and these estimates are also upper bounds* although they are weaker bounds than those in (1.10) and (1.7).

We will return to this in later chapters. Before we go into more details on data structures and other important features and applications of triangulation, we describe a simple and useful algorithm for constructing triangulations.

1.4 A Triangulation Algorithm

Various algorithms for triangulation are described in detail in later chapters and in particular in Chapter 4. Here we will work out a fairly straightforward algorithm for establishing a triangulation from a set of points surrounded by a closed polygon. Given a set of points in the plane without any closed polygon surrounding the point set, we can always construct such a polygon by selecting a number of points outside the data set. There are many ways to do this, however the most common way is to select four points as corners of a rectangle so that all given points are contained in the rectangle. There also exist methods for computing the convex hull of a point set, which is the smallest convex region in the plane containing the points. However, as we shall see later, the convex hull of a set of points can be established by using triangulation methods.

To make things somewhat more interesting than just embedding the points into a rectangle, we will present the algorithm by using a closed polygon as illustrated in Figure 1.7.

Note that the closed polygon must be a simple closed polygon, which basically means that the polygon must not intersect itself and must satisfy the condition labelled (7) in Section 1.2. The algorithm has three major steps:

- Triangulation of the closed polygon without considering the interior points.
- Insert the interior points, one by one, keeping the triangulation updated as a regular triangulation for each insertion.

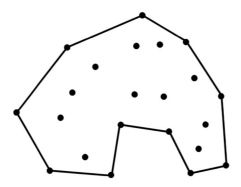

Fig. 1.7. Closed polygon surrounding a point set.

- Improve the triangulation by edge-swapping – *improving* a triangulation and *edge-swapping* will be explained later.

Triangulation of a closed polygon. The first step is to subdivide the interior of the closed polygon into triangles. Figure 1.8(a) shows a closed simple polygon $\mathcal{P}$. We first define what we mean by a protruding point.

Definition 1.2 (Protruding point). *A point p_i of a closed simple polygon $\mathcal{P}$ is called a protruding point if the following conditions hold. (Figure 1.8(a))*

1. *The interior angle at p_i of the triangle $t_{i-1,i,i+1}$ is less than π. We use the convention $p_{N+1} = p_1$, where N is the number of points in $\mathcal{P}$, and $t_{i-1,i,i+1}$ is the triangle given by the three points p_{i-1}, p_i and p_{i+1}.*
2. *The triangle $t_{i-1,i,i+1}$ contains no other points of $\mathcal{P}$ apart from p_{i-1}, p_i, and p_{i+1}.*

It is obvious that any simple closed polygon must have at least one protruding point. An algorithm for triangulating the interior of a simple closed polygon $\mathcal{P}$ with $N > 3$ vertices, can be as follows.

Algorithm 1.1 Protruding point removal

---

1. Find an arbitrary protruding point p_i of $\mathcal{P}$. (Figure 1.8(a))
2. Join the points p_{i-1} and p_{i+1} to form an edge $e_{i-1,i+1}$ and create the triangle $t_{i-1,i,i+1}$.
   Let $\mathcal{P}$ be the new polygon after excluding p_i from the list of points in the old polygon. (Figure 1.8(b)
3. **if** $\mathcal{P}$ has only three points, i.e. if $\mathcal{P}$ is a triangle,
      stop. (Figure 1.8(f))
4. **goto** 1.

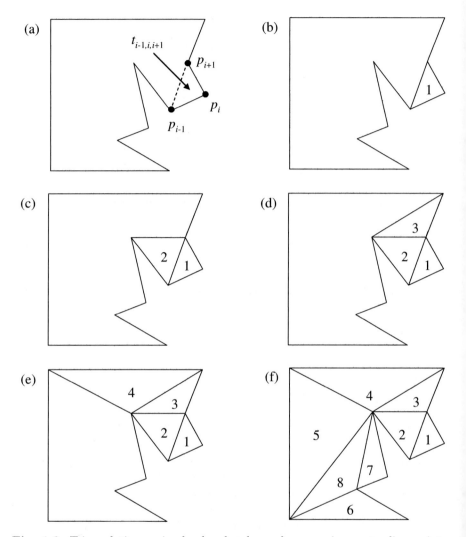

Fig. 1.8. Triangulating a simple closed polygon by removing protruding points successively.

By applying this algorithm, we can triangulate a closed polygon as shown in Figure 1.7, and obtain a result as in Figure 1.9. Note that the result is not unique.

Point insertion. After triangulating the closed polygon $\mathcal{P}$ surrounding the points, the next step is to insert the interior points into the triangulation. A set of distinct interior points $\{p_i\}$ can be inserted successively by Algorithm 1.2,

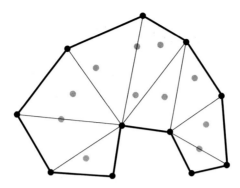

Fig. 1.9. Triangulation of the closed polygon in Figure 1.7.

where we denote the interior of a triangle t_j by $Int(t_j)$, and the interior of an edge e_k by $Int(e_k)$.

Algorithm 1.2 Point insertion

---

1. Locate the triangle t_j which contains p_i, (t_j is some triangle in the triangulation of $\mathcal{P}$)
2. **if** $p_i \in Int(t_j)$
    replace t_j with three new triangles as shown in Figure 1.10(a).
    else if $p_i \in Int(e_k)$, where e_k is an edge of t_j,
    replace t_j and the other triangle sharing e_k with t_j, with four triangles as shown in 1.10(b).

We observe that the number of triangles and edges in the triangulation is increased by two and three respectively for each point insertion, and that these numbers are independent of whether $p_i \in Int(t_j)$ or $p_i \in Int(e_k)$. This agrees with the formulas given in (1.2) and (1.3).

If the above point insertion method is used on our example in Figures 1.7 and 1.9, we obtain a triangulation of the whole point set as shown in Figure 1.11. The result depends on the order in which points are inserted, hence this point insertion method does not give a unique triangulation. The figure shows two triangulations where the points are inserted in different orders.

An issue of importance in the practical use of triangulations is to handle numerical problems which arise when a point for insertion is very close to existing points and edges. Given a distance measure $dist(\cdot, \cdot)$, we may introduce a tolerance ε and apply the following rules:

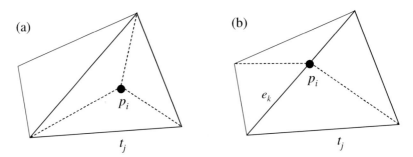

Fig. 1.10. Inserting a point p_i in a triangulation. In (a), p_i is in the interior of a triangle, and in (b), p_i is in the interior of an edge.

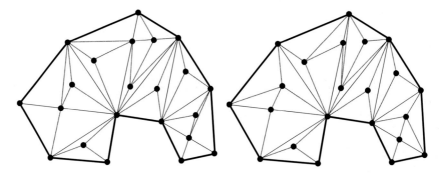

Fig. 1.11. Two different triangulations of the same point set produced by the point insertion algorithm.

- If a point p is such that $dist(p, p_i) \leq \varepsilon$, the point p is not inserted.
- If $dist(p, e_{i,j}) \leq \varepsilon$, insert the point as though it was on the edge, (the old edge is replaced by two edges which is not a split of the old edge).

As already indicated and illustrated in Figure 1.11, a triangulation constructed using the algorithms above is not unique. Another problem is that usually triangles constructed with these algorithms will vary considerably in shape. In particular we obtain long and skinny triangles. In other words, we encounter considerable variation in angle size.

Edge-swapping. The third step of the algorithm is to improve the triangulation. Here we may decide that a "good triangulation" is one with as few small interior angles as possible. Since the positions of the vertices are kept fixed, we must define new edges between vertices. One way to do this is to swap edges which are diagonals in *strictly convex quadrilaterals*. A quadrilateral is a closed polygon with four edges, and we say that it is strictly convex if all its four interior angles are less than $180°$. In Figure 1.12 the edge $e_{i,i+2}$ is the diagonal of the strictly convex quadrilateral with vertices v_i, v_{i+1}, v_{i+2} and v_{i+3}. The edge $e_{i,i+2}$ can be swapped to become a new edge $e_{i+1,i+3}$

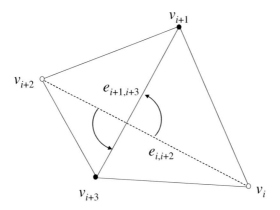

Fig. 1.12. Swapping an edge in a strictly convex quadrilateral. The dotted line indicates the edge before swapping.

in the triangulation. This swapping operation is illustrated by the arrows in Figure 1.12. Thus, two triangles are replaced with two new triangles. We can now describe an algorithm based on this technique.

Algorithm 1.3 Edge-swapping

---

1. Locate the diagonal of a strictly convex quadrilateral.
2. Swap the diagonal if the smallest of the six interior angles within the quadrilateral increases. In Figure 1.12 we observe that a swap from $e_{i,i+2}$ to $e_{i+1,i+3}$ will be preferred since the smallest angle is larger when using $e_{i+1,i+3}$ rather than $e_{i,i+2}$.
3. **if** all strictly convex quadrilaterals have been checked and no swaps have been performed,
   > **stop**
   else
   > **goto** Step 1.

Applying the algorithm to our example in Figure 1.11 we obtain, after a few swaps, the result in Figure 1.13. We may start the edge-swapping with any of the two triangulations in Figure 1.11, or any other triangulation of the same point set, but we will always obtain the result in Figure 1.13. The explanation for this and more details on the algorithm are given in Chapter 3.

A sequence of such edge-swaps is performed to optimize a triangulation according to an optimization criterion. For example, we observe that some of the triangles in Figure 1.11 are elongated and almost degenerate. In many applications we want to avoid such "poorly shaped" triangles since they may lead to oscillating surfaces or numerical problems in calculations.

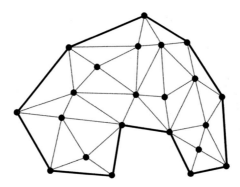

Fig. 1.13. Final triangulation when the edge-swapping of Algorithm 1.3 has terminated.

Finally we want to mention here that all possible triangulations of a set of vertices, given a boundary of the triangulation, can be reached by a sequence of edge-swaps starting from an initial triangulation. The number of possible triangulations is, of course, finite. In Chapter 4 we present more triangulation algorithms that produce so-called Delaunay triangulations.

1.5 Edge Insertion

Predefined edges in triangulations are frequently used for representing rivers and roads in terrain models and geological faults in geological horizon modeling. Predefined edges are commonly referred to as *constrained edges*, and a triangulation with predefined edges is called a *constrained triangulation*. Chapter 6 is devoted entirely to this important construction.

Figure 1.14 shows an initial regular triangulation Δ, and an edge $e_{i,j}$ between two points p_i and p_j. We want to insert the edge into the triangulation such that the resulting triangulation is valid and regular according to the definitions in Section 1.2. The endpoints p_i and p_j of $e_{i,j}$ have already been inserted, for example using Algorithm 1.2. We start by removing all edges that are intersected by $e_{i,j}$, or more precisely all triangles that are intersected by $e_{i,j}$, but retaining the edges of these triangles that are not intersected by the edge. The removed triangles define a region, often referred to as the influence region of $e_{i,j}$. The next step is to triangulate the influence region such that the constrained edge is included in the set of edges in the triangulation. A sequence of edges, for example a polygon representing a river, can be inserted in an existing triangulation by inserting each edge of the polygon successively as in Algorithm 1.4.

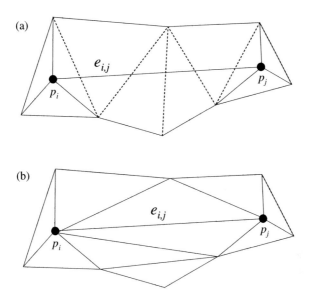

Fig. 1.14. Inserting an edge $e_{i,j}$ in a triangulation. The dotted lines in (a) shows which edges are removed. The retriangulated region is shown in (b).

Algorithm 1.4 Constrained edge insertion

---

1. **for each** triangle t_i of Δ,
       if $Int(t_i) \cap e_{i,j} \neq \phi$
       remove t_i from Δ.
   We obtain one or more regions R_i, with simple closed polygons as boundaries, on each side of $e_{i,j}$ that are not subdivided into triangles. More than one region on each side can occur if $e_{i,j}$ intersects triangles at their vertices. In Figure 1.14 we obtain two regions, one on each side of $e_{i,j}$.
2. Triangulate each region R_i using Algorithm 1.1.

Note that, since the number of boundary vertices $|V_B|$ is fixed for each region R_i, we can determine the number of triangles and edges in the triangulation directly from the basic formulas introduced in Lemma 1.1.

A topology structure with adjacency information between vertices, edges and triangles is needed to avoid a worst-case execution time for Algorithm 1.2 and 1.4. For example, in Step 1 of Algorithm 1.4 we will need to search through the whole list of triangles in Δ if a suitable topology structure is not present. We will return to topology and data structures in more detail in Chapter 2. As for point insertion, we must take care of numerical problems that may occur when the insertion edge is close to one or more of the existing points of the triangulation.

1.6 Using Triangulations

We conclude this introductory chapter with some comments on the use of triangles and triangulations. In later chapters we will go into more detail on a few important applications based on triangulations. Among the huge number of applications using triangulation, we will focus on construction and representation of surfaces, mesh generation for finite element calculations and triangle-based methods for real-time visualization of huge data sets. In particular we will focus on triangulation of geographical data.

From the history of surveying and cartography we find that triangulation has been used as the basic technique for calculating the positions of landmarks in the terrain. Assume that three positions are known in some appropriate coordinate system, and for simplicity we use Cartesian coordinates. We have three points on the terrain with known positions, e.g. mountain peaks a, b and c. Visible from the three mountain peaks is another mountain peak, say d. By measuring horizontal and vertical angles, the position of d can be calculated from a, b and c and the measured angles. Continuing this process, more and more points on the Earth's surface can be determined. Today, equipment for measuring position on the Earth's surface is more sophisticated, for example the global position system (GPS).

A surface in 3D space can be constructed by assigning height values to each vertex of a triangulation as indicated in Figure 1.15. The surface is defined as a continuous piecewise linear surface, that is, planar facets over each triangle joined together without gaps.

Figure 1.16 shows a terrain model where the underlying triangulation was constructed by the algorithms in the preceding sections. We have also included sequences of constrained edges (polygons representing roads and rivers) in the terrain model. We will call these constructions *surface triangulations*.

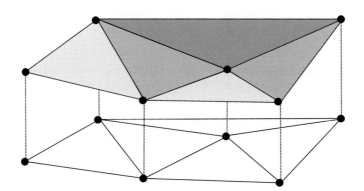

Fig. 1.15. Triangulation and linear surface patches.

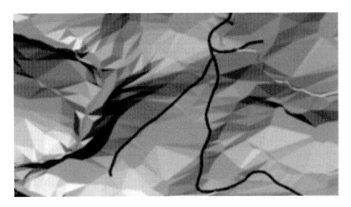

Fig. 1.16. A terrain surface with constrained edges – roads and rivers.

In mathematical terms a surface triangulation f defined over a triangulation Δ with $|T|$ triangles belongs to the finite dimensional function space $S_1^0(\Delta)$ of piecewise linear polynomials,

$$S_1^0(\Delta) = \left\{ f \in C^0(\Omega) : f|_{t_i} \in \Pi_1, \; i = 1, \ldots, |T| \right\}.$$

Here Π_1 is the space of bivariate linear polynomials to which the restriction of f on a triangle t_i of Δ belongs. Further, Ω is the domain triangulated by Δ, and $C^0(\Omega)$ is the space of "0-continuous functions" in Ω that are continuous, but their derivatives are not continuous across the triangle edges of Δ. S_1^0 takes its '1' from Π_1 and its '0' from C^0.

Let $z_1, \ldots, z_{|V|}$ be the height values assigned to the $|V|$ vertices of the triangulation. Then the interpolation conditions $f(x_i, y_i) = z_i$, for $i = 1, \ldots, |V|$, determine a unique surface triangulation in $S_1^0(\Delta)$. If the height values assigned to vertices p_1, p_2 and p_3 of a triangle t are z_1, z_2 and z_3 respectively, then the height value of the surface at an arbitrary point p inside t is

$$z_p = b_1 z_1 + b_2 z_2 + b_3 z_3, \tag{1.12}$$

where (b_1, b_2, b_3) are the barycentric coordinates of p with respect to p_1, p_2 and p_3. Thus, to evaluate the function f at a point p, one must

i) locate the triangle t containing p,
ii) calculate the barycentric coordinates of p with respect to the vertices of t by Equation (1.1), and then
iii) apply Equation (1.12) to find $f(p) = z_p$.

In geographic information systems and geological modeling systems, piecewise linear surfaces in $S_1^0(\Delta)$ are often sufficient and may also be preferred to higher-degree surfaces, which may cause undesirable oscillatory behavior (often referred to as Gibbs phenomenon) when modeling terrain with rapidly varying topography. Piecewise linear surfaces are also more efficient to compute and more compliant with other software components in such applications.

In other applications, one may prefer surface triangulations that are smoother than those that belong to $S_1^0(\Delta)$. A popular choice is the so-called Clough-Tocher interpolant for constructing surface triangulations f in the space $S_3^1(\Delta_{CT})$ [18]. Here Δ_{CT} is a refined triangulation obtained by splitting each triangle of Δ in its barycenter $(b_1, b_2, b_3) = (1/3, 1/3, 1/3)$ and thus replacing each triangle in Δ by three new triangles. The origin of the method is from finite element analysis. In addition to assigning height values to each vertex of Δ, one must also supply gradients (normal vectors) at the vertices, or the gradients may be estimated based on the piecewise linear interpolant in $S_1^0(\Delta)$ to the given data. The restriction of f to each triangle in Δ_{CT} is now a bivariate cubic (degree three) polynomial that can be represented on the so-called Bernstein-Bezier form. The surface patches defined over each triangle are joined smoothly together with continuous first partial derivatives $\partial f / \partial x$ and $\partial f / \partial y$ across the triangle edges.

Triangulations are otherwise used for representing subdivisions of general two-manifold surfaces, opened or closed. Thus, a point p_i in the three dimensional Euclidean space is associated with each vertex v_i of the triangulation.

1.7 Exercises

1. Calculate the centers of the circumcircles (circumcenters) for the points
   a) $(0, 1)$, $(1, 1)$ and $(0, 1)$
   b) $(0, 0)$, $(0, 4)$ and $(8, 4)$
   c) (a, b), $(2a, b)$ and $(4a, 2b)$
   d) (a, b), (b, a) and $(-a, -b)$
2. For what values of a, b, c and d do the three points fall on one straight line?
   a) (a, b), $(2a, b)$ and $(4a, 2b)$
   b) (a, b), (c, d) and $(a + c, b + d)$
   c) (a, b), (c, d) and $(a + b, c + d)$
3. Prove that the three bisecting lines in Figure 1.2 intersect at one point c.
4. Calculate the smallest angle of each of the triangles given by the points in Exercise 1.
5. Draw triangulations with at least five nodes such that the number of triangles and edges equals the lower bounds and upper bounds given by the inequalities in (1.6) and (1.7).
6. How many triangles and edges result from triangulation of a simple closed polygon with N vertices?

7. Insert an edge between two existing nodes in a triangulation. How many edges and triangles are there in the new triangulation compared to the old?

8. Show that equations (1.3) and (1.4) hold for a regular triangulation.

9. Prove Lemma 1.1 by successively adding new triangles instead of removing triangles as in the existing proof.

10. Prove (1.6) and (1.7).

11. Generalize equations (1.2), (1.3) and (1.4) for:

   a) the triangulation contains n holes.
   b) the surface triangulation is closed, that is, $|E_B| = 0$.

2

Graphs and Data Structures

The *topology* of a triangulation can be described by graph theoretic concepts such that a clear distinction is made between the topological structure and the geometric embedding information. The topological elements of a triangulation are nodes (or vertices), edges and triangles, and the geometric embedding information, which is associated with these elements, is points, curves (or straight-line segments) and surface patches respectively. Likewise, a distinction can be made between topological and geometric *operators*. By considering triangulations as *planar graphs*, we can benefit from an extensive theory and a variety of interesting algorithms operating on graphs. In particular, we will see that *generalized maps*, or *G-maps*, provide useful algebraic tools to consider triangulations at an abstract level. Common data structures for representing triangulations on computers are outlined and compared in view of storage requirements and efficiency of carrying out topological operations.

2.1 Graph Theoretic Concepts

A *graph* $G(V, E)$ consists of a set of *vertices* V and a set E of pairs of vertices in V called *edges*. Figure 2.1 shows geometric representations of two graphs where vertices are drawn as circles, and edges are drawn as straight-line segments connecting pairs of vertices. In general, vertices of a graph can be isolated without belonging to pairs of vertices constituting edges. The trivial graph is a single vertex.

It is evident that a triangulation Δ as defined in Chapter 1 can be considered as a graph if adjacent triangles share a common edge. We will use the notation $G_\Delta(V, E)$ for the graph representation of a triangulation that consists of a set of vertices V and a set of edges E, thus the set of triangles is implicitly represented in the graph as *cycles* where each cycle consists of three vertices and three edges.

(a)   (b)

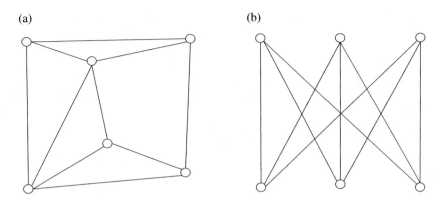

Fig. 2.1. A planar embedding of a graph in (a), and a non-planar graph in (b).

We will work mostly with triangulations that represent tessellations of manifold surfaces. If the surface is not closed the triangulation will have a boundary and the graph can be restricted to be a planar graph.

Definition 2.1 (Planar Graph). *A planar graph G is a graph that can be embedded (or drawn) in the plane such that the embedding of an edge e_{ij} in the plane intersects G only at the vertices v_i and v_j.*

The graph shown in Figure 2.1(a) is planar according to this definition, and the graph shown in Figure 2.1(b) can be proved to be *non-planar*. If the edges of a planar graph can be embedded in the plane as straight-line segments, such as the graph in Figure 2.1(a), we also call it a *planar straight-line graph (PSLG)*, also referred to as a planar subdivision or a map by some authors.

If the pairs of vertices in the set E are *ordered*, that is, if directions are imposed on the edges in $E = \{(v_i, v_j)\}$, we call G a *directed graph* or a *digraph*. Otherwise, if we allow for permutations of the vertices v_i and v_j in the set of edges, we call G an *undirected graph*.

If a graph G does allow more than one edge connecting a pair of vertices, we call G a *multigraph*. Otherwise, if a graph does not allow parallel edges, we call it a *simple graph*. A graph that is both planar and directed in addition to being a multigraph, will be called a *planar directed multigraph*.

A graph $G(V, E)$ is said to be *connected* if there is a path between any two vertices in V along the edges in E, otherwise the graph is *disconnected*.

There are several common generalizations of graphs and many special graphs that can be deduced from the definitions above. For example, in the next section we will introduce a graph notation for triangulations that also includes relationship information between the topological entities, that is, between vertices, edges and triangles.

In the context of triangulations, we must choose a graph representation and a corresponding data structure that is capable of reflecting the relationships between topological entities needed by algorithms. The general definition of a simple graph $G(V, E)$, without directions imposed on the edges in E, definitely yields a valid graph representation, but it is not always adequate. An efficient data structure based on half-edges, which will be introduced later, will consist of two directed edges between pairs of vertices. Thus, the graph representation will be a planar directed multigraph as explained above.

There is plenty of literature on graphs, algorithmic graph theory and applications of graphs. The interested reader is referred to [88, 57, 58] and corresponding references.

2.2 Generalized Maps (G-maps)

There are conceptually two different classes of topological operators required for triangulations: operators for traversing the topology structure (topological queries), and operators that change the topology (topological modifiers). As far as the former is concerned, there are a number of possibilities and needs depending on the actual data structure and application in mind. Basically, however, it is possible to define a few simple elementary operators that all other traversal operators can be composed of. We will define these operators using concepts of *darts* and *involutions* in generalized maps, or G-maps. G-maps can also be used to define topological modifiers, but we leave this discussion for Chapter 9.

G-maps, are general tools for modeling the topology of boundary-based geometric models (B-reps) and have been applied in application areas such as geological modeling [67, 74, 82, 39]. G-maps are algebraically defined based on a limited number of clear concepts. The topology is described using a single topological element, the *dart* , and a set of functions operating on the set of darts in the topology structure. In this section, G-maps are applied to obtain an algebraic description of the topology of triangulations, and a level of abstraction at which topological operations can be defined. For a thorough theoretical discussion of G-maps, the reader is referred to [53] and [10].

A dart in a triangulation structure can be considered as a unique triple $d = (v_i, e_j, t_k)$, where v_i is a node (or a vertex) of the edge e_j, and v_i and e_j are a node and an edge of the triangle t_k. Thus, for each triangle there are six possible combinations of the triple that can define a dart. In Figure 2.2(a), a dart d is indicated both as an arrow and as a (node, edge, triangle)triple. Referring to Figure 2.2(b), we define three unique functions, α_0, α_1 and α_2 operating on the set of triples D in a triangulation as one-to-one mappings from D onto itself,

$$\alpha_i : D \to D, \ i = 0, 1, 2.$$

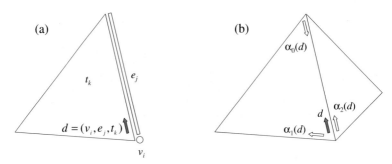

Fig. 2.2. (a): A dart considered as a unique triple $d = (v_i, e_j, t_k)$. (b): Shows how d is mapped under the functions α_i, $i = 0, 1, 2$.

In the theory of generalized maps, these functions are called involutions as they are bijections with the property $\alpha_i(\alpha_i(d)) = d$. We will simply call them α-*iterators* and define them as follows when applied to triangulations:

- $\alpha_0(d)$ maps d to a triple with a different node, but keeps the edge and the triangle fixed,
- $\alpha_1(d)$ maps d to a triple with a different edge, but keeps the node and the triangle fixed,
- $\alpha_2(d)$ maps d to a triple with a different triangle, but keeps the node and the edge fixed.

That is, $\alpha_0(d)$, $\alpha_1(d)$ and $\alpha_2(d)$ change the node, edge and triangle of the triple $d = (v_i, e_j, t_k)$ respectively, while the other topological elements are kept fixed. If an edge e_j in a triple $d = (v_i, e_j, t_k)$ is at the boundary of the triangulation, then $\alpha_2(d) = d$, hence the node, the edge and the triangle are all kept fixed. This is called a *fixed point* of the α_2-iterator. With the definitions above, a triangulation can be considered as a *G-map*,

$$G(D, \alpha_0, \alpha_1, \alpha_2),$$

which defines a combinatorial structure, or an algebra, on the set of darts D.

More informally, one can think of a dart $d = (v_i, e_j, t_k)$ as an element positioned inside the triangle t_k at the node v_i and at the edge e_j as depicted in Figure 2.2(a). The dart changes its content and position in the triangulation structure according to the definition of α_i, $i = 0, 1, 2$ operating on d. The α_0 and α_1 iterators "reposition" the dart inside the same triangle while α_2 "moves" the dart over an edge to the neighboring triangle t_k.

We will also say that a dart has a clockwise or counterclockwise direction with respect to a triangle as seen from one side of the triangulation. Thus, the dart d in Figure 2.2(a) is oriented counterclockwise in the triangle t_k as determined by the direction of the arrow symbolizing the dart. Note that $\alpha_i(d)$, $i = 0, 1, 2$ changes the direction of d from clockwise to counterclockwise,

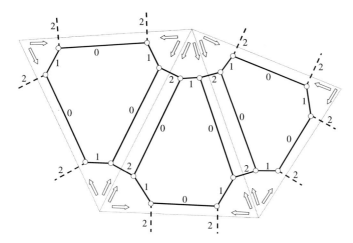

Fig. 2.3. A triangulation with three triangles, and its corresponding labelled graph defined as the dual of the G-map.

or vice versa, except when α_2 is applied to a dart positioned at an edge at the boundary of the triangulation. We also define the regions "*left of d*" and "*right of d*" in the plane as seen from one side of the triangle and in the direction of the arrow. Lastly, a dart $d = (v_i, e_j, t_k)$ can also be interpreted as a vector with direction from v_i to the opposite node of v_i, that is, to the node v_i' associated with the triple of the dart $d' = (v_i', e_j, t_k) = \alpha_0(d)$.

A G-map, and thus a triangulation, has a corresponding labelled graph that can be considered as the dual of the G-map [54]. The triangulation in Figure 2.3 consists of three triangles drawn with dotted edges, whereas the labelled graph is drawn with bold edges. Each node of the graph corresponds to a dart drawn as an arrow close to the node and pointing in the direction of the node. The labelled edges incident with[1] each node correspond to the α-iterators applied to the dart that correspond to that node. For example, an edge labelled "1" links two nodes of the graph corresponding to two darts d_i and d_j that are linked through the α_1-iterator; $\alpha_1(d_i) = d_j$ and $\alpha_1(d_j) = d_i$. The dotted edges labelled "2" correspond to fixed points of the α_2-iterator at the boundary of the triangulation. The graph is *regular of degree three* in the sense that every node has exactly three edges incident with it.

In the following, we denote a *composition* $\alpha_i(\alpha_j(\cdots \alpha_k(d) \cdots))$ by $\alpha_i \circ \alpha_j \circ \cdots \circ \alpha_k(d)$. Note that a composition $\alpha_i \circ \alpha_j(d)$, with $i \neq j$, does not change the orientation of d inside a triangle from clockwise to counterclockwise, or vice versa, unless the composition involves $\alpha_2(d)$ where d is at the boundary of the triangulation. Note also the following interesting properties that we will use later when composing iterators operating on triangulations.

---

[1] We say that an edge with nodes v_i and v_j as endpoints is *incident with* v_i and v_j.

- Applying the composition $\alpha_0 \circ \alpha_1$ repeatedly iterates through the nodes and the edges of a triangle.
- Applying the composition $\alpha_1 \circ \alpha_2$ repeatedly iterates through all edges and triangles sharing a common node.
- Applying the composition $\alpha_0 \circ \alpha_2$ repeatedly iterates around an edge by visiting the two nodes of the edge and the two triangles sharing the edge.
- If the order of a composition $\alpha_i \circ \alpha_j$ is swapped to $\alpha_j \circ \alpha_i$, the iteration goes in the opposite direction.

In addition we have,

$$\alpha_i \circ \alpha_i(d) = d, \ i = 0, 1, 2, \ \text{and}$$
$$(\alpha_0 \circ \alpha_2(d))^2 = \alpha_0 \circ \alpha_2 \circ \alpha_0 \circ \alpha_2(d) = d.$$

Definition 2.2 (Orbit and k-orbit, $k = 0, 1, 2$). *Let $\{\alpha_i\}$ be one, two or all three α-iterators of a G-map $G(D, \alpha_0, \alpha_1, \alpha_2)$ and let $d \in D$. An orbit $\langle\{\alpha_i\}\rangle(d)$ of a dart d is the set of all darts in D that can be reached by successively applying compositions of α-iterators in $\{\alpha_i\}$ in any order starting from d. The k-orbit of a dart d in a G-map is defined as the orbit $\langle\alpha_i, \alpha_j, \ i, j \neq k, \ i \neq j\rangle(d)$.*

Thus, the 0-orbit, $\langle\alpha_1, \alpha_2\rangle(d)$, and the 1-orbit, $\langle\alpha_0, \alpha_2\rangle(d)$, is the set of all darts around a node and around an edge respectively. The 2-orbit, $\langle\alpha_0, \alpha_1\rangle(d)$, is the set of all darts inside a triangle, see Figure 2.4. We observe that the orbit $\langle\alpha_0, \alpha_1, \alpha_2\rangle(d)$, d being any dart of D, is the set of all darts in D provided that all triangles in the triangulation associated with G are connected edge by edge. Thus, all nodes, edges and triangles, of a triangulation can be reached by compositions of α_0, α_1 and α_2 starting from any dart $d = (v_i, e_j, t_k)$.

The concept of darts and α-iterators can be generalized to arbitrary dimensions n such that for an n-G-map we have iterators α_i, $i = 0, \ldots, n$. For example, for $n = 3$ the topology of tetrahedrizations can be modelled by

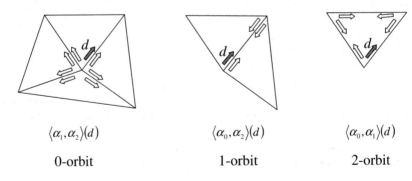

$$\langle\alpha_1, \alpha_2\rangle(d) \qquad \langle\alpha_0, \alpha_2\rangle(d) \qquad \langle\alpha_0, \alpha_1\rangle(d)$$

0-orbit   1-orbit   2-orbit

Fig. 2.4. The k-orbits, $k = 0, 1, 2$, of a dart d in a G-map.

associating a dart with a quadruple, $(node, edge, triangle, tetrahedron)$, and iterators α_i, $i = 0, 1, 2, 3$, and compositions of them can be used for navigating in the topology structure as similarly described above for triangulations. Also, arbitrary boundary-based models can be modelled by G-maps; that is, faces need not be triangles and volumes need not be tetrahedra.

The α-iterators are basically the only traversal operators needed on a triangulation. Thus they can be used as an interface to arbitrary data structures, and generic algorithms for navigating in the topology can be based solely on these iterators. By means of compositions $\alpha_i \circ \alpha_j \circ \cdots \circ \alpha_k(d)$, it is then possible to build any traversal operator needed by an application. In Chapter 9 we specify in detail how this can be implemented generically, using function templates in the C++ programming language.

2.3 Data Structures for Triangulations

There are many possible topological structures, or data structures, for representing triangulations on computers. A data structure must be chosen in view of the needs and requirements in the actual application. When analyzing different data structures one is always faced with a trade-off between storage requirements and efficiency of carrying out topological and geometric operations. For example, for visualization purposes one needs a data structure with fast access to data and sufficient topological information for traversing the topology fast when extracting sequences of triangles for the visualization system. This will normally require more storage than a data structure used only for storing a triangulation in a database. In a real application, one might need more than one data structure and tools for mapping one to the other.

Figure 2.5(a) shows a triangulation with seven nodes and six triangles only where nodes and triangles are labeled. We use the notation introduced in Section 1.2, where a triangle $t_{i,j,k}$, which spans the three vertices v_i, v_j and v_k, is represented as an ordered triple (i, j, k). The vertices are always referenced in counterclockwise order as seen from one side of the triangulation. For example, the triangle labeled '1', in the figure can be represented as the triple $(1, 7, 2)$.

When estimating storage requirements for data structures, we assume that the number of vertices is large and that the number of vertices at the boundary is much smaller than the total number of vertices in the triangulation. Recall the estimates derived in Section 1.3 for the number of triangles and edges, and the degree sum of the vertices in a triangulation under these assumptions,

$$|T| \approx 2|V|, \qquad |E| \approx 3|V|, \qquad \sum_{i=1}^{|V|} \deg(V_i) = 2|E| \approx 6|V|,$$

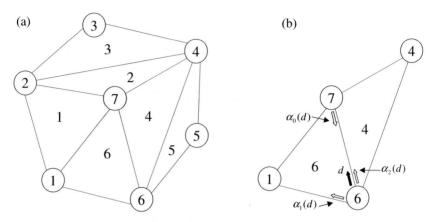

Fig. 2.5. A sample triangulation is shown in (a), and (b) shows how α_i, $i = 0, 1, 2$ repositions a dart d.

where $|V|$, $|E|$ and $|T|$ are the number of vertices, edges and triangles in a triangulation respectively. We also derived lower and upper bounds for the number of triangles and edges. We will use the upper bounds

$$|T| \leq 2|V| - 5 \text{ and}$$
$$|E| \leq 3|V| - 6$$

when analyzing storage requirements for the data structures discussed in the next sections.

The geometric positions of the vertices, in 2D or 3D, must always be stored. If a real number requires 8 bytes on a computer, $24|V|$ bytes are required for storing the positions of the vertices in 3D space, or $16|V|$ bytes in 2D space. If the triangulation represents a terrain model and the terrain surface is represented as triangular planar patches, the geometric positions of the vertices are usually the only geometric embedding information needed. In other applications, such as CAD (computer aided design), one might also need other embedding information, for example coefficients of Bezier patches when modeling smooth surfaces [28].

We will also briefly comment on the complexity of carrying out topological traversal operations in data structures in terms of the α-iterators. Figure 2.5(b) shows a dart d positioned in one of the triangles in the sample triangulation and it is shown how d is repositioned under α_i, $i = 0, 1, 2$. Apart from compact storing and fast traversal operations, a data structure must also enable efficient updating of its topology.

2.4 A Minimal Triangle-Based Data Structure

A minimal data structure, where minimal refers to the amount of storage used to represent a triangulation, is simply to store the triples $I_\Delta = \{(i,j,k)\}$ representing the set of triangles in the triangulation in a list or an array. The triangulation in Figure 2.5(a) can be stored thus:

Triangle #	Triangle		
	i	j	k
1:	1	7	2
2:	2	7	4
3:	2	4	3
4:	7	6	4
5:	4	6	5
6:	7	1	6

The triangles are stored in arbitrary order, and a triple (i,j,k), which represents a triangle that spans v_i, v_j and v_k, has no unique starting point, though the ordering of the vertices is counterclockwise. In C++, the fields in the columns i, j and k would be represented as pointers to node objects or as integers representing node numbers. In the following we will call this *pointer fields*. The storage requirement in terms of the number of pointer fields, N_P, is now three pointers per triangle. Using the upper bound, $|T| \le 2|V| - 5$, and the estimate $|T| \approx 2|V|$, gives

$$N_P \le 6|V| - 15 \text{ and}$$
$$N_P \approx 6|V|.$$

If each pointer field is stored as a four-byte integer, $24|V|$ bytes is required for storing the table above. This is exactly the same as the storage requirement for the geometric positions of the vertices in 3D space.

There is no adjacency information between triangles in this representation, so topological operators are costly and tedious to implement. The α_0- and α_1-iterators defined in Section 2.2 for navigating inside a triangle are straightforward to implement and fast to execute, but the α_2-iterator, which is needed for switching from one triangle to another (Figure 2.5(b)), requires a search in $O(|T|)$ time in the list of triangles, where $|T|$ is the number of triangles.

One could use this data structure for visualization purposes by displaying one triangle at a time, but this would not allow the use of optimized structures for fast rendering, such as triangle strips and triangle fans, which are common in graphics rendering systems, see Section 2.9. However, the data structure

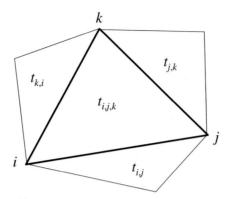

Fig. 2.6. A triangle $t_{i,j,k}$ and adjacent triangles $t_{j,k}$, $t_{k,i}$ and $t_{i,j}$ opposite vertices i, j and k respectively.

can be used for storing a triangulation compactly in a database together with algorithms for mapping it to a more efficient topological structure in the memory of the computer.

2.5 Triangle-Based Data Structure with Neighbors

A natural extension of the data structure above is to introduce adjacency information such that for each triangle $t_{i,j,k}$ the three triangles sharing an edge with $t_{i,j,k}$ are recorded. We use the notation $t_{i,j}$ when referring to a neighboring triangle that has $e_{i,j}$ as an edge, i.e., which joins the two vertices v_i and v_j, see Figure 2.6.

Two lists are needed, each of the same size: one triangle list identical to the list in the previous section and a neighbor list with three references for each triangle to the neighboring triangles. The topological representation of the triangulation in Figure 2.5(a) will now be represented as shown in the table,

Triangle #	Triangle i	j	k	neighbors $t_{j,k}$	$t_{k,i}$	$t_{i,j}$
1:	1	7	2	2	-	6
2:	2	7	4	4	3	1
3:	2	4	3	-	-	2
4:	7	6	4	5	2	6
5:	4	6	5	-	-	4
6:	7	1	6	-	4	1

Each triple of triangles in a neighbor record is listed counterclockwise. $t_{j,k}$ is the neighbor triangle opposite vertex v_i in the triangle $t_{i,j,k}$, and correspondingly for $t_{k,i}$ and $t_{i,j}$. A hyphen indicates the boundary of the triangulation. The storage requirement for the topological information is twice as much as for the minimal data structure above. We get $N_P = 6|T|$ and thus,

$$N_P \le 12|V| - 30 \text{ and}$$
$$N_P \approx 12|V|$$

in number of pointer fields when using the upper bound $|T| \le 2|V| - 5$ and the estimate $|T| \approx 2|V|$.

This data structure contains topological relationships between triangles as opposed to the minimal data structure in Section 2.4. Given a triangle $t_{i,j,k}$ we know exactly what triangles are adjacent to it and where they are positioned relative to the vertices and the edges of $t_{i,j,k}$. The α_0- and α_1-iterators have the same complexity as in the minimal data structure, but α_2 can be carried out in $O(1)$ operations (constant time) as opposed to $O(|T|)$ for the minimal structure.

2.6 Vertex-Based Data Structure with Neighbors

A storage-efficient data structure based on vertices only was introduced by Cline & Renka [17]. For each vertex v_i in a triangulation, the vertices which are joined to v_i by an edge are recorded in an adjacency list. If v_i is at the boundary of the triangulation, the adjacency list is terminated with a "pseudo-node". Since the number of neighbors is not the same for each vertex, one must also keep track of the number of neighbors for each vertex. Figure 2.7 shows a boundary vertex of the sample triangulation in Figure 2.5(a) and its adjacent vertices. The table below shows how the whole triangulation can be represented.

Vertex #	Adjacent vertices (ADJ)						End list (END)
1	6	7	2	0			4
2	1	7	4	3	0		9
3	2	4	0				12
4	3	2	7	6	5	0	18
5	4	6	0				21
6	5	4	7	1	0		26
7	6	4	2	1			30

The *ADJ* lists contain the adjacent vertices for each vertex v_i in counterclockwise order. If v_i is on the boundary of the triangulation, the listed

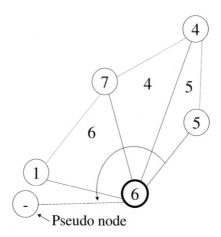

Fig. 2.7. A vertex labelled '6' at the boundary of a triangulation and its adjacent vertices. The pseudo-node indicates the boundary.

neighbors are terminated with a zero (pseudo-node). The entries in the END list represent, for each vertex v_i, the accumulated number of neighbors for the previous vertices and v_i itself, including pseudo-nodes. If one-dimensional arrays ADJ and END are used for storing the integers in the table, then the neighbors of vertex v_i, for $v_i > 1$, begin in $ADJ(END(v_i - 1) + 1)$ and v_i has $END(v_i) - END(v_i - 1)$ neighbors (including possibly a pseudo-node).

The storage requirement in number of pointer fields for ADJ and END is $N_{ADJ} = \sum_{i=1}^{|V|} \deg(V_i) + |V_B|$, including the pseudo-nodes, and $N_{END} = |V|$. Using (1.8) and (1.3) we obtain

$$
\begin{aligned}
N_{ADJ} &= 2|E| + |V_B| \\
&= 2(3|V| - |V_B| - 3) + |V_B| \\
&= 6|V| - 2|V_B| - 6 + |V_B| \\
&= 6|V| - |V_B| - 6 \\
&\leq 6|V| - 9,
\end{aligned}
$$

where the inequality is obtained from the lower bound, $3 \leq |V_B|$, for the number of boundary vertices. The total number of pointer fields is,

$$
\begin{aligned}
N_P &= N_{ADJ} + N_{END} \\
&= 7|V| - |V_B| - 6 \\
&\approx 7|V|
\end{aligned}
$$

when $|V|$ is large compared to the number of boundary vertices $|V_B|$. This is considerable less than the triangle-based data structure in the previous section.

In some applications, it is useful to have a list of triangles, although triangles are not part of the actual data structure. A naive representation of triangles similar to that of the minimal triangle-based data structure in Section 2.4 would require approximately $6|V|$ additional pointer fields. Thus the total storage requirement would be $N_p \approx 13|V|$ if triangles are recorded, and this is slightly more than the triangle-based structure. The basic topological traversal operations α_0, α_1 and α_2 have complexity $O(1)$ and are straightforward to implement.

2.7 Half-Edge Data Structure

The notion of *half-edge* as the basic topological entity for boundary-based topological representations was introduced by Weiler [87]. The principle is to split each edge into two directed half-edges, each of which are oriented opposite the other. Hence, we can think of a half-edge as belonging to exactly one triangle. The three half-edges of a triangle can be oriented counterclockwise around the triangle as seen from one side of the triangulation. A half-edge has its *source node* where it starts from, and its *target node* where it points to. Figure 2.8(a) shows the half-edge representation of the triangulation used previously in this chapter. Note that this data structure reflects a planar directed multigraph as defined in Section 2.1.

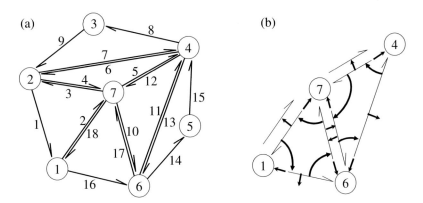

Fig. 2.8. Half-edge data structure.

There are many possible data structures that can be derived from this concept, for example, the *vertex-edge* and the *face-edge* data structure [87]. A minimal pointer structure, where minimal refers to minimal storage while maintaining "sufficient" information for topological operators, is shown in Figure 2.8(b). Each half-edge has a pointer to the vertex it starts from, a pointer to the next half-edge belonging to the same triangle (counterclockwise) and a pointer to its "twin-edge" belonging to another triangle. The table below shows the pointer fields for the whole sample triangulation. A hyphen in the rightmost column indicates that the half-edge has no twin-edge, thus the half-edge is at the boundary of the triangulation.

Half-edge #	Vertex	Half-edge pointers	
		Next edge in triangle	Twin-edge
1:	2	2	-
2:	1	3	18
3:	7	1	4
4:	2	5	3
5:	7	6	12
6:	4	4	7
7:	2	8	6
8:	4	9	-
9:	3	7	-
10:	7	11	17
11:	6	12	13
12:	4	10	5
13:	4	14	11
14:	6	15	-
15:	5	13	-
16:	1	17	-
17:	6	18	10
18:	7	16	2

Let N_V be the number of vertex pointers and N_{E_h} the number of half-edge pointers in the data structure. Since half-edges at the boundary have no twin-edge, $|E_h| < 2|E|$, and using the upper bound $|E| \leq 3|V| - 6$ gives,

$$N_V < 6|V| - 12 \text{ and}$$
$$N_{E_h} < 12|V| - 24.$$

The total storage requirement in number of pointer fields for the half-edge data structure is

$$N_P = N_V + N_{E_h} \approx 18|V|.$$

If, for a certain application, we also need to maintain a list of triangle pointers, we could represent each triangle as one of its half-edges. This would require approximately $2|V|$ additional pointers and the total storage requirement would be $N_P \approx 20|V|$. This is considerably more than the triangle-based and the vertex-based data structures in the previous sections, but, as can be easily verified, the α-iterators can be executed faster with this data structure. An example of object-oriented design for the half-edge data structure can be found in Chapter 9.

2.8 Dart-Based Data Structure

The dart as the basic topological element of the data structure yields extremely fast topological traversal operations. This concept has been applied in geological fault network modeling [39, 38]. The data structure can be represented as a set of darts D where each dart $d \in D$ has references to a vertex and to three other darts in D as defined by the iterators α_0, α_1 and α_2. Thus, each of the basic topological operations can be implemented simply as an entry in an array, or as a pointer from one dart object to another dart object when using an object-oriented programming language such as C++. Figure 2.9(a) shows the model example used above with the set of darts D, and in (b) the pointer structure for two of the triangles is shown. Each dart is given a number positioned at the start of the arrow symbolizing the dart. Note that the darts numbered 1 and 2 in (b) have references to themselves. This is in accordance with the definition of a fixed point at the boundary of the triangulation where $\alpha_2(d) = d$.

The table below shows the pointer fields for the darts of the two triangles in Figure 2.9(b). There are four pointer fields for each dart $d = (v_i, e_j, t_k)$; a reference to the vertex v_i of d, and references to the three darts resulting from applying $\alpha_0(d)$, $\alpha_1(d)$ and $\alpha_2(d)$.

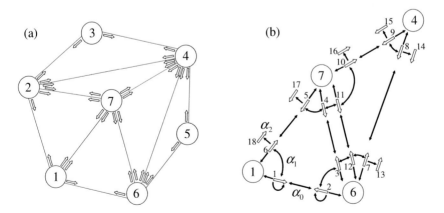

Fig. 2.9. Dart-based data structure.

Dart d	Dart pointers			
	Vertex	$\alpha_0(d)$	$\alpha_1(d)$	$\alpha_2(d)$
1:	1	2	6	1
2:	6	1	3	2
3:	6	4	2	12
4:	7	3	5	11
5:	7	6	4	17
6:	1	5	1	18
7:	6	8	12	13
8:	4	7	9	14
9:	4	10	8	15
10:	7	9	11	16
11:	7	12	10	4
12:	6	11	7	3

The storage requirement is indeed considerably more than any of the data structures described above. There are six darts for each triangle, and each dart has four references. Using the upper bound $|T| \leq 2|V| - 5$ and the estimate $|T| \approx 2|V|$ for the number of triangles, we get for the number of pointer fields N_P,

$$N_P \leq 48|V| - 120 \text{ and}$$
$$N_P \approx 48|V|.$$

As for the half-edge data structure, a list of pointers to triangles could also be preferable. If one of the six darts (a "leading dart") of a triangle represents the triangle, then we would need approximately $2|V|$ additional pointer fields and the total storage requirement would be $N_P \approx 50|V|$ for the dart-based data structure. This is almost three times the storage requirement for the half-edge data structure. However, though we pay a high price for storage, we gain efficiency in terms of a fast execution of topological traversal operations. The α-iterators are carried out instantly without any of the tests or calculations which were necessary with other data structures.

2.9 Triangles for Visualization

Triangle strips and *triangle fans* are constructions used for fast renderings of triangulations. Instead of delivering one triangle each time to the graphics processor, sequences of triangles can be delivered in chunks. This is supported by common graphics hardware such that the number of lower-level operations in the graphics pipeline can be minimized when visualizing large triangulations where many triangles share the same vertices. Such performance gains

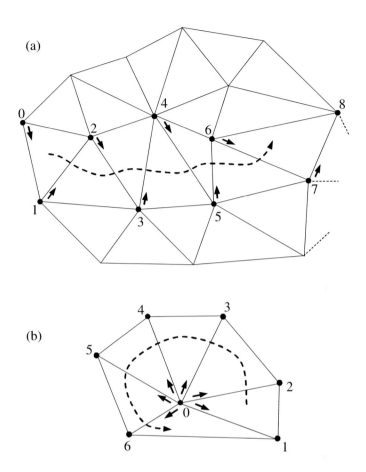

Fig. 2.10. Triangle strip in (a) and triangle fan in (b).

are decisive for large-scale terrain visualization, in scientific visualization, in virtual reality environments and when rendering large polygonal models in computer games.

Consider the triangulation in Figure 2.10(a), and the sequence Δ_s consisting of $|T_s| = 7$ triangles intersected by the dotted arrow. By using vertex indices, this part of the triangulation can be represented triangle by triangle by the simple triangle-based data structure in Section 2.4,

$$[0, 1, 2], [2, 1, 3], [2, 3, 4], [4, 3, 5], [4, 5, 6], [6, 5, 7], [6, 7, 8]. \qquad (2.1)$$

But Δ_s can also be considered as a sequence of adjacent triangles ordered regularly such that each pair of adjacent triangles share a common edge. The triangles can be traversed by visiting all vertices in the order in which they are numbered in the figure starting from vertex 0 and following a regular zig-zag

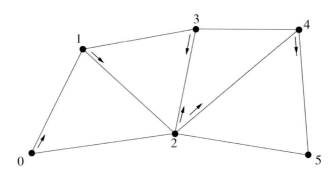

Fig. 2.11. Triangle strip with a degenerate triangle $[2, 3, 2]$.

pattern along the edges as indicated by the arrows in the figure. Thus, the shorter sequence of vertices

$$\{0, 1, 2, 3, 4, 5, 6, 7, 8\}$$

also gives a unique representation of Δ_s in Figure 2.10(a). This triangle strip representation keeps the first and the last triple of the sequence in (2.1) and the last index in each of the intermediate triples. The first three indices define a triangle and then each additional index defines another triangle by using the two preceding indices. In general, the number of indices of a triangle strip is only $|T_s| + 2$, compared to $3|T_s|$ indices for an array representing the triangulation triangle by triangle. A triangle strip also implicitly holds adjacency information between triangles.

In some situations it is necessary to repete a vertex to get a legal encoding of a triangle strip. In order to represent the triangulation in Figure 2.11 as one contiguous triangle strip, we must repete the vertex labelled 2, and we get the sequence

$$\{0, 1, 2, 3, 2, 4, 5\}.$$

Here $[2, 3, 2]$ represents a degenerate triangle. Strips with degenerate triangles are properly handled by common graphics hardware.

A triangle fan is a sequence of adjacent triangles sharing a common vertex. The triangle fan in Figure 2.10(b) can be represented uniquely by the sequence

$$\{0, 1, 2, 3, 4, 5, 6, 1\}.$$

From a G-map point of view a triangle fan is a 0-orbit. Thus, the composition $\alpha_1 \circ \alpha_2(d)$ can be used to rotate around the vertex with index 0 from triangle to triangle, collecting the sequence of vertex indices representing the triangle fan. Similarly, applying the composition $\alpha_2 \circ \alpha_1 \circ \alpha_0(d)$ one can traverse Δ_s in Figure 2.10(a) to collect vertex indices representing the triangle strip.

OpenGL, which is the standard API (Application Programming Interface) for graphics rendering in 3D applications [1, 89], offers triangle strips and triangle fans as geometric primitives for efficient rendering of triangulations. The following code fragment in C++ shows how the triangle strip in Figure 2.10(a) can be delivered to OpenGL:

```
// Define vertex geometry (coordinates) and assign indices
GLfloat vtx_array[]={< x1,y1,z1,x2,y2,z2,x3,y3,z3,...>};
glVertexPointer(3, GL_FLOAT, 9, 0, vtx_array);
GLuint idx_array[]={0, 1, 2, 3, 4, 5, 6, 7, 8};

// Draw the triangle strip
glDrawElements(GL_TRIANGLE_STRIP, 9,
               GL_UNSIGNED_INT,
               idx_array);
```

Similarly, the triangle fan in Figure 2.10(b) can be delivered to OpenGL with the C++ code:

```
GLfloat vtx_array[]={< x1,y1,z1,x2,y2,z2,x3,y3,z3,...>};
glVertexPointer(3, GL_FLOAT, 7, 0, vtx_array);
GLuint idx_array[]={0, 1, 2, 3, 4, 5, 6, 1};
glDrawElements(GL_TRIANGLE_FAN, 8,
               GL_UNSIGNED_INT,
               idx_array);
```

There are many interesting theoretical results, which have practical implications, on computing triangle strips (see, for example, [6] and [27]). Moreover, there are many strategies for obtaining "optimal" decompositions into strips and fans, but what is "optimal" regarding the number of triangles in each sequence and the organization of the sequences depends on the graphics hardware. Scene graphs, such as the OpenSceneGraph [2], have utility functions for creating triangle strips and triangle fans from triangulations and other polygonal data.

2.10 Binary Triangulations

In this section we outline an efficient and elegant data structure for representing a special type of triangulation. The triangulation is *semi-regular* in the sense that the triangle vertices constitute a subset of a regular grid. The data structure is derived from a scheme known as *longest edge bisection* that has become popular for view-dependent visualization [55, 72]. In the literature

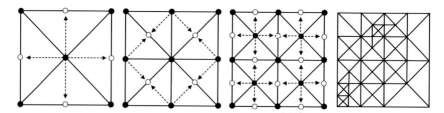

Fig. 2.12. Recursive longest edge bisection. Dotted arrows indicate parent-child relationships. The rightmost triangulation has only three active vertices from the finest level.

the scheme is also called $4 - k$ meshes [84], or restricted quad-tree triangulations [66]. Here we call meshes generated by this scheme *binary triangulations* since they can be considered as the result of recursive splitting of one triangle into two new triangles.

Consider the initial triangulation Δ_a on the left in Figure 2.12 with four isosceles triangles. Recursive splitting of Δ_a starts by inserting a vertex on the longest edge of each triangle. We call the bisected edge the *split-edge* and the inserted vertex the *split-vertex*. In order to maintain a valid triangulation (cf. Section 1.2), four new edges must be introduced as indicated by dotted arrows. Each arrow represents a parent-child relationship, where the black bullet represents the *parent* and each circle represents a *child*. The new triangulation Δ_b (second from left) has eight isosceles triangles whose longest edges are bisected to obtain Δ_c with 16 triangles (second from right). The dotted arrows in Δ_c indicate parent-child relationships involved when generating the next triangulation in the recursion. Each interior vertex in any triangulation Δ_k at level k has exactly two parents in Δ_{k-1}, while a vertex at the boundary (except the four corner vertices) has only one parent. We say that a vertex is *active* when it has been inserted into the mesh. The corner vertices i_{sw}, i_{se}, i_{nw} and i_{ne} as well as the center vertex i_c are always active by convention. (See Figure 2.13).

If every longest edge at every level is bisected, the vertices at any level k constitute a regular grid, and the grid at the next finer level $k + 1$ is obtained by inserting grid lines halfway between the grid lines at level k. However, every vertex of the regular grid at a level need not be active for the triangulation to be valid. An example is shown by the rightmost triangulation in Figure 2.12, where only three vertices are active at the finest level. The general rule is that if a vertex at a certain level is active, then its parents, grandparents, and all its ancestors at coarser levels are also active. This prepares the ground for *adaptive* meshes defined by binary triangulations. In Chapter 8 we utilize this powerful feature in a multilevel scheme for approximating huge scattered data sets over triangulations.

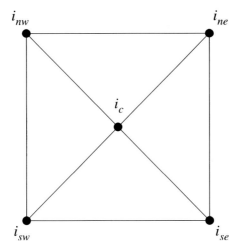

Fig. 2.13. Vertices and triangles at the coarsest level. The four corner vertices and the center vertex are always active.

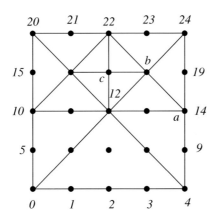

Fig. 2.14. Numbering of vertices in a binary triangulation with two levels.

A simple numbering of the vertices of the binary triangulation reveals beautiful properties of algorithms operating on these triangulations. Let the vertex indices at the finest level be numbered linearly along the rows as in Figure 2.14, where the number of levels $n = 2$. We observe that the number of vertices at the finest level along any of the four boundary segments of the domain is equal to $2^{n/2} + 1$. The index of a vertex v_i can be found as the average

$$i = \frac{l + r}{2},$$

where v_l and v_r are vertices spaced equally long from v_i along a horizontally, vertically or diagonally line. For example, v_l and v_r can be vertices of the split-edge corresponding to the split vertex v_i, or if v_i is an interior vertex, v_l and v_r can be the parents of v_i. In Figure 2.14 we see that vertex a has index $(4+24)/2 = 14$, b has index $(12+24)/2 = 18$ and c has index $(12+22)/2 = 17$.

A data structure for representing binary triangulations is straightforward to implement. In fact, the triangulation can be represented implicitly only by its (active) vertices and by references from each vertex to its two parents. In many applications, like view-dependent visualization as presented by Lindstrom & Pascucci in [56], the binary triangulation will be dynamic with vertices that alternate between being active and non-active. In this case all vertices at the finest grid must be stored with a field that indicates whether the vertex is active or not. The ordering of the vertices should correspond to the numbering in Figure 2.14.

Algorithms operating on the data structure become simple and efficient, for example algorithms for locating triangles and algorithms for generating triangle strips for efficient visualization. We illustrate this by a simple example. Suppose we want to traverse a binary triangulation and draw all triangles one at a time. Let the four corner vertices and the vertex in the middle be numbered as in Figure 2.13. The following two functions implements this functionality by an elegant recursion.

```
MeshRefine
    SubmeshRefine(i_sw, i_c, i_se, 1)
    SubmeshRefine(i_se, i_c, i_ne, 1)
    SubmeshRefine(i_ne, i_c, i_nw, 1)
    SubmeshRefine(i_nw, i_c, i_sw, 1)
```

```
SubmeshRefine(i_l, i_a, i_r, l)
    j := ½(i_l + i_r)
    if (l < n & IsActive(j))         // n is the number of levels
        SubmeshRefine(i_a, j, i_l, l + 1)
        SubmeshRefine(i_r, j, i_a, l + 1)
    else
        DrawTriangle(i_l, i_r, i_a)
```

A simple recursion similar to that above can also be used to traverse the triangulation and make triangle strips. If we accept degenerate triangles, as discussed in Section 2.9, we can make one contiguous triangle strip of the whole triangulation. (See Figure 2.15 and Exercise 6.)

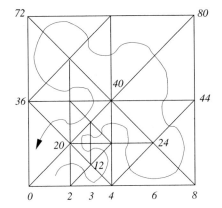

Fig. 2.15. Triangle strip extraction from a binary triangulation with three levels. Example by Lindstrom & Pascucci [56].

2.11 Exercises

1. Given a triangulation represented by the vertex-based data structure in Section 2.6, write an efficient algorithm for generating a list of triangles where each triangle is represented by three nodes in counterclockwise ordering (as in Section 2.4).

2. Write a program that converts from the vertex-based data structure in Section 2.6 to the triangle-based data structure in Section 2.5 (for example by extending the algorithm from the exercise above).

3. Write programs that extract the boundary from a triangulation represented by
   a) the triangle-based data structure in Section 2.5,
   b) the vertex-based data structure in Section 2.6.

4. Decompose the rest of the triangulation in Figure 2.10 into triangle strips and triangle fans.

5. Find the sequence of vertex indices that represents the triangle strip in Figure 2.15.

6. Write pseudo code (similar to `MeshRefine` and `SubmeshRefine` in Section 2.10) that extracts a binary triangulation as one contiguous triangle strip. (See Figure 2.15.)

7. Find an algorithm for locating the triangle in a binary triangulation that contains the point (u, v) in the plane.

3

Delaunay Triangulations and Voronoi Diagrams

This chapter is devoted to two important constructs in computational geometry, the Voronoi diagram and the Delaunay triangulation. These constructions are dual in the sense that one can be defined, or derived, from the other. They have been used in a number of applications, and the theory has been studied over many years and is well understood. In the following sections, we present a unified discussion of the classical theory of Voronoi diagrams and Delaunay triangulations in the plane, and thus provide a theoretical basis for designing algorithms for Delaunay triangulation.

3.1 Optimal Triangulations

In Chapter 1 we introduced a simple and straightforward method for constructing triangulations over polygonal regions in the plane. We observed that the triangulations were not necessarily unique, as a sequence of edge-swaps could be applied to obtain new valid triangulations. In many applications, we wish to optimize a triangulation in some sense to avoid "poorly shaped" triangles, such as triangles that are elongated or almost degenerate. This has given rise to many criteria for deciding what makes a triangualtion "good", and a variety of algorithms have been developed that strive for optimality according to these criteria.

Motivated by avoiding poorly shaped triangles, we may say that a triangulation is "good" if it consists of triangles that are close to being equiangular. More specifically, if we compare all possible triangulations constructed from the same point set, we may prefer one that has a triangle with the *smallest maximal* angle, or alternatively, we may prefer one with a triangle that has the *largest minimal* angle. These criteria are known as the *MinMax* and the *MaxMin* angle criterion respectively.

Figure 3.1 shows two different triangulations Δ^a and Δ^b of the same point set P. The domain in which both triangulations are defined is the convex hull

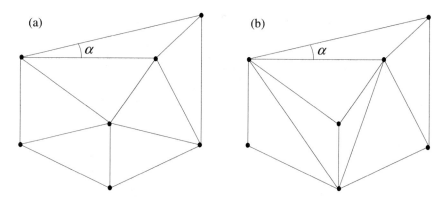

Fig. 3.1. Two triangulations of the same point set that satisfy the MaxMin angle criterion. The triangulation in (a) is a Delaunay triangulation.

of P. The triangulation Δ^b is obtained from Δ^a by swapping two of the edges in Δ^a. The angle α shown in both triangulations is the largest minimal angle of all possible triangulations of P, and thus, both Δ^a and Δ^b satisfy the MaxMin angle criterion. We observe that Δ^b has more poorly shaped triangles than Δ^a in the sense that there are more small interior angles in Δ^b. Following the arguments above, though we still do not have a precise criterion for an optimal triangulation, we may prefer Δ^a to Δ^b in some applications even though Δ^b also satisfies the MaxMin angle criterion. Later in this chapter we will explain how the triangulation in Figure 3.1(a) is constructed and why we can claim that the angle α is the largest minimal angle of all possible triangulations of P.

We now make a more precise measure of "goodness" of a triangulation as mentioned above, and introduce criteria for optimal triangulations in the sense of the MaxMin and the MinMax angle criteria. Assume that for each possible triangulation Δ^k of a point set P we associate an *indicator vector* $I(\Delta^k) = (\alpha_1, \alpha_2, \ldots, \alpha_{|T|})$ where $|T|$ is the number of triangles in Δ^k. If we assume that all triangulations have the same boundary, then the number of triangles is fixed, cf. (1.2). In the case of the MaxMin angle criterion, each entry α_i represents the smallest interior angle of each triangle in Δ^k and the entries of each indicator vector are arranged in non-decreasing order,

$$I(\Delta^k) = (\alpha_1, \alpha_2, \ldots, \alpha_{|T|}), \ \alpha_i \leq \alpha_j, \ i < j.$$

An ordering of all vectors $I(\Delta^k)$ will impose a linear ordering on the set of all possible triangulations of P. A vector $I(\Delta^i)$ which is *lexicographically larger*[1] than a vector $I(\Delta^j)$, which we will denote by $I(\Delta^i) > I(\Delta^j)$, may now

---

[1] We say that a vector $\mathbf{v}^1$ is lexicographically larger than a vector $\mathbf{v}^2$ if for some integer m, we have $v_i^1 = v_i^2$ for $i = 1, \ldots, m-1$ while $v_m^1 > v_m^2$.

represent a triangulation Δ^i that is "better" than a triangulation Δ^j, (cf. the discussion of the triangulations in Figure 3.1 above). When $I(\Delta^i) > I(\Delta^j)$ we say that Δ^i strictly follows Δ^j in the linear ordering. Further, we define the *optimal triangulation* in the MaxMin sense as the triangulation corresponding to the vector I which is lexicographically largest, that is, no triangulation follows it in the linear ordering. The triangulations in Figure 3.1(a) and (b) have indicator vectors

$$I(\Delta^a) = (14.04,\ 26.57,\ 36.87,\ 40.60,\ 40.60,\ 49.40,\ 50.91) \text{ and}$$
$$I(\Delta^b) = (14.04,\ 15.26,\ 19.44,\ 26.57,\ 29.74,\ 36.87,\ 45.00).$$

We observe that the first entry of both vectors are identical, but the second entry in $I(\Delta^a)$ is larger than the second entry in $I(\Delta^b)$. That is, $I(\Delta^a)$ is lexicographically larger than $I(\Delta^b)$, so the triangulation in Figure 3.1(a) is "more optimal" than the triangulation in (b). In fact, Δ^a is an optimal triangulation in the MaxMin sense since it is constructed with an algorithm that is guaranteed to find an optimal triangulation. We will return to this later.

In the case of the MinMax angle criterion, the entries of the vectors represent the largest interior angle of each triangle and the entries β_i of the vectors are arranged in the opposite order, i.e., non-increasing so that

$$J(\Delta^k) = (\beta_1, \beta_2, \ldots, \beta_{|T|}),\ \beta_i \geq \beta_j,\ i < j.$$

We define the *optimal triangulation* in the MinMax sense as the triangulation corresponding to the vector J which is lexicographically smallest. One possibility is to use this criterion when creating triangulations. However, we leave the MinMax angle criterion here and concentrate on the MaxMin angle criterion that is used below to define the *Delaunay triangulation*.

Since the number of possible triangulations of a point set is finite, it is obvious that there always exists an optimal triangulation for both the MaxMin angle criterion and the MinMax angle criterion. But there may be more than one optimal triangulation. We return to this in the next section when we define the so-called neutral case.

Based on the MaxMin angle criterion, in which case the smallest angle is maximized, we are now in a position to give one of many equivalent definitions of a Delaunay triangulation:

Definition 3.1 (Delaunay triangulation, MaxMin angle criterion). *A triangulation that is optimal in the sense of the MaxMin angle criterion and which is defined on the convex hull of a point set is called a Delaunay triangulation.*

This particular type of triangulation has been widely studied in the literature. It has many interesting characteristics and there is extensive mathematical theory we can benefit from. The Delaunay triangulation is also fairly easy to compute. This is in contrast to other types of triangulations for which theory merely exists and which are difficult to compute. The triangulation Δ^a in Figure 3.1 is a Delaunay triangulation.

The name Delaunay triangulation is attributed to the Russian matematician Boris Delaunay [22]. The correct name is Boris Nikolaevich Delone (1890–1980), but Delone is pronounced "Delaunay" and now also commonly spelled that way in English.

In the reminder of this chapter we will study different geometric characteristics of Delaunay triangulations and arrive at other definitions which will be shown to be equivalent to the definition above. Algorithms for the construction of Delaunay triangulations are studied in Chapter 4. But first we define the neutral case, which is important to have in mind when developing the theory of Delaunay triangulations further.

3.2 The Neutral Case

Consider the vertices of a rectangle. These four points are cocircular meaning that there is a unique circle passing through the points. There exist two possible triangulations of the points depending on which diagonal of the rectangle that is chosen as an edge in the triangulation. One triangulation can be obtained from the other by swapping the diagonal of the rectangle. Exactly the same angles occur in both triangulations. Thus, both the largest minimal angle and the smallest maximal angle occur in both triangulations, and therefore both satisfy the MaxMin angle criterion and the MinMax angle criterion defined above.

A configuration of four cocircular points with two possible triangulations that both satisfy a certain criterion, is referred to as a *neutral case* for that criterion. Although the theory of Delaunay triangulations can handle this, we often assume that neutral cases for the MaxMin angle criterion do not exist when developing the theory. This leaves out lengthy details and boring explanations in the proofs of theorems.

Another example of a neutral case, but now only neutral with respect to the MaxMin angle criterion, is shown in Figure 3.2 with triangulations Δ^a on the left and Δ^b on the right. The vertices of the triangulations are cocircular and form a strictly convex quadrilateral. The indicator vectors with respect to the MaxMin angle criterion are

$$I(\Delta^a) = (\alpha_1^a, \alpha_2^a) \quad \text{and}$$
$$I(\Delta^b) = (\alpha_1^b, \alpha_2^b).$$

Thus, α_1^a and α_1^b are the smallest angles in Δ^a and Δ^b respectively. But α_1^a and α_1^b span the same arc on the circumscribed circle, so these two angles are equal. Since the smallest angle occurs in both triangulations, both satisfy

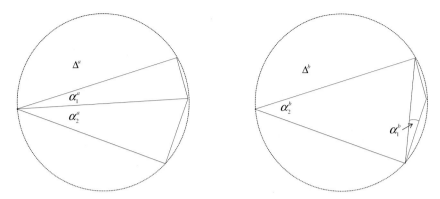

Fig. 3.2. A neutral case for the MaxMin angle criterion. Only Δ^a satisfy the Min-Max criterion. The triangulation Δ^b is optimal with respect to the MaxMin angle criterion.

the MaxMin angle criterion. Therefore this is a neutral case for the MaxMin angle criterion. On the other hand, the largest angle only occurs in Δ^b on the right, so this is not a neutral case for the MinMax angle criterion - only the triangulation Δ^a on the left, which has the smallest maximal angle, satisfies the MinMax angle criterion.

A neutral case for the MaxMin angle criterion occurs if and only if the four points of the convex quadrilateral are cocircular. It can be shown that for a neutral case for the MinMax angle criterion to occur, two adjacent angles must be equal and they must be the maximum angles in the quadrilateral [40].

When examining the second element of the indicator vectors, we find that $\alpha_2^b > \alpha_2^a$. Therefore $I(\Delta^b)$ is lexicographically larger than $I(\Delta^a)$, so only Δ^b is *optimal* with respect to the MaxMin angle criterion although both satisfy the MaxMin angle criterion. According to Definition 3.1, Δ^b is a Delaunay triangulation. In the literature, Δ^a is often also referred to as a Delaunay triangulation, even though it is not optimal with respect to the MaxMin angle criterion. We will also see that Δ^a satisfies two definitions of a Delaunay triangulation given later in this chapter. Therefore, we try to avoid these ambiguities in the following by often assuming that neutral cases do not exist.

3.3 Voronoi Diagrams

In this section we introduce the concept of Voronoi diagrams which have close relations to Delaunay triangulations. Consider a set of distinct points $P = \{p_1, \ldots, p_N\}$ in the plane and let $d(p_i, p_j)$ denote the Euclidean distance between p_i and p_j. To each point $p_i \in P$ we associate a region in the plane

$$V(p_i) = \{x \; : \; d(x, p_i) < d(x, p_j), \; j = 1, \ldots, N\}, \tag{3.1}$$

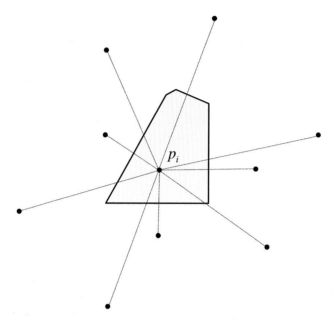

Fig. 3.3. The Voronoi region of a point p_i in the plane.

which we call the *Voronoi region* of p_i. That is, $V(p_i)$ consists of all points x in the plane that are closer to p_i than to any other point in P. Figure 3.3 shows a set of points and the Voronoi region associated with a point p_i. The "construction" of $V(p_i)$ can be done as follows. Let $H(p_i, p_j)$ denote the half-plane containing all points in the plane that are closer to p_i than to p_j. Thus, $H(p_i, p_j)$ contains p_i and is defined by the perpendicular bisector of the line between p_i and p_j. ($H(p_i, p_j)$ is the Voronoi region of p_i with respect to the point set $\{p_i, p_j\}$.) Then the Voronoi region $V(p_i)$ is the intersection of $N - 1$ half-planes,

$$V(p_i) = \bigcap_{\substack{j=1,\ldots,N \\ j \neq i}} H(p_i, p_j), \qquad (3.2)$$

where each Voronoi region has at most $N - 1$ sides.

If we apply this construction to all points in P, we obtain a unique set of non-intersecting Voronoi regions $V(p_i)$, $i = 1, \ldots, N$ as shown in Figure 3.4. If we denote by $\overline{V(p_i)}$ the closure of $V(p_i)$, that is,

$$\overline{V(p_i)} = \{x \; : \; d(x, p_i) \leq d(x, p_j), \; j = 1, \ldots, N\},$$

then the union of all $\overline{V(p_i)}$, $i = 1, \ldots, N$ covers the whole plane.

From the definitions above, it is clear that all Voronoi regions are open. In [68], it is shown that Voronoi regions associated with the points on the

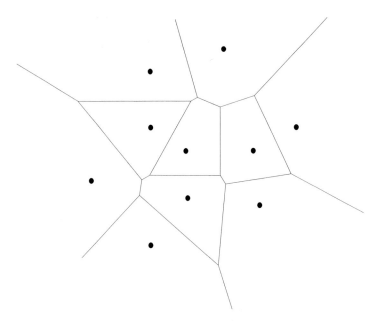

Fig. 3.4. The Voronoi diagram of a set of points in the plane.

convex hull of P are unbounded and that the other regions are bounded. Since a half-plane is a convex region and, by (3.2), a Voronoi region is the intersection of half-planes, the Voronoi regions are also convex. The collection of all Voronoi regions is called the *Voronoi diagram*, or the *Dirichlet tessellation*, of the given point set P. Further, the boundaries of the Voronoi regions are often referred to as *Voronoi polygons*, the edges of the polygons as *Voronoi edges* and the vertices of the Voronoi polygons as *Voronoi points*. Two points p_k and p_l in P are said to be *Voronoi neighbors* if their associated Voronoi regions $V(p_k)$ and $V(p_l)$ share a common Voronoi edge. The closest point to a point p in a point set is always a Voronoi neighbor of p, while the other Voronoi neighbors of p depend on the distribution of P in the plane. (Exercise 6.)

Voronoi diagrams have been used in many scientific disciplines for structuring adjacency information between irregularly distributed points. For example, the climatologist Thiessen used Voronoi regions in connection to interpolation of climatic data from unevenly distributed weather stations [83]. Because of this, the regions $V(p_i)$ are also called *Thiessen regions*.

3.4 Delaunay Triangulation as the Dual of the Voronoi Diagram

The Voronoi diagram defined in the previous section leads directly to a constructive geometric definition of a Delaunay triangulation. These constructions are dual in the sense that one can be defined, or derived, from the other. In the discussion below we leave out some of the details and proofs of our claims regarding the Voronoi diagram and its dual, the Delaunay triangulation. For in-depth theory, analysis and completeness, the interested reader is referred to literature on computational geometry, such as the book by Preparata & Shamos [68] and references given there.

Recall the definition of Voronoi neighbors above and suppose that we connect all Voronoi neighbors by straight-line segments as shown in Figure 3.5. It can be shown that this construction always results in a subdivision of the plane into a collection of non-overlapping adjacent triangles covering the convex hull of P. The line segments connecting Voronoi neighbors associated with unbounded Voronoi regions define the convex hull of P, and the other line segments define common edges of pairs of adjacent triangles. With this construction we have reached another definition of a Delaunay triangulation in addition to the definition in Section 3.1.

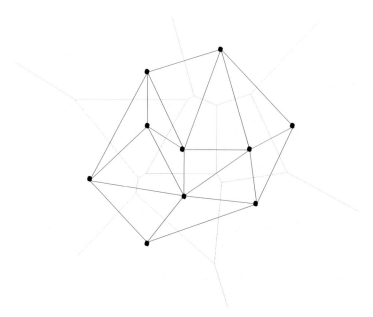

Fig. 3.5. The straight line dual of the Voronoi diagram which yields a Delaunay triangulation.

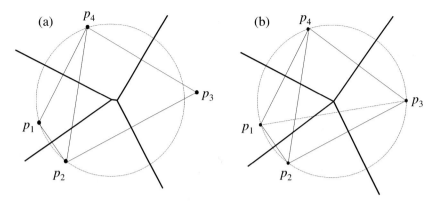

Fig. 3.6. Almost cocircular points p_1, p_2, p_3 and p_4 left, and cocircular points right. In the cocircular case there are two possible solutions when constructing the Delaunay triangulation.

Definition 3.2 (Delaunay triangulation, straight line dual). *A Delaunay triangulation is the straight-line dual of the Voronoi diagram.*

The triangles obtained from the construction above are called *Delaunay triangles*. Similarly, the edges that connect Voronoi neighbors in P are called *Delaunay edges*. We will show later that the collection of Delaunay triangles obtained thus also constitutes a Delaunay triangulation as given by Definition 3.1. Another useful geometric definition of a Delaunay triangulation, equivalent to the two definitions already given, will come shortly.

Note that while the Voronoi diagram is uniquely defined by the construction in Section 3.3, a Delaunay triangulation obtained from the Voronoi diagram is not necessarily unique. Consider a triangulation of four points that are almost cocircular as in Figure 3.6(a). The points p_2 and p_4 are Voronoi neighbors and they are connected by an edge in the Delaunay triangulation. If we move the point p_3 to the circumcircle of triangle $t_{1,2,4}$ as shown in Figure 3.6(b), one of the Voronoi edges degenerates to a point such that p_1 and p_3 also become Voronoi neighbors. When connecting Voronoi neighbors to form edges of the Delaunay triangulation, there are two possible solutions as depicted in the figure. Similarly, if more than four points are cocircular, we can complete the triangulation by connecting points on the circle arbitrarily to form edges of the triangulation such that edges do not intersect except at their endpoints. The neutral case, where four or more points in P are cocircular, is the only case where any ambiguity can arise. Otherwise, if no more than three points are cocircular, the Delaunay triangulation of a finite set of points is *unique*.

From the constructive definition above we also conclude, without a formal proof, the *existence* of a Delaunay triangulation of a finite set of points P.

In a neutral case as in Figure 3.6(b), two points that can be connected with an edge in a Delaunay triangulation are called *weak Voronoi neighbors*. Thus, p_1 and p_3 are weak Voronoi neighbors and also p_2 and p_4 are weak Voronoi neighbors. Otherwise, in non-neutral cases, when two points can be connected to form an edge in a Delaunay triangulation they are called *strong Voronoi neighbors*. For example, in the non-neutral case in Figure 3.6(a), p_2 and p_4 are strong Voronoi neighbors (and so are the other points that form edges in the Delaunay triangulation). In both cases we call the edges connecting Voronoi neighbors Delaunay edges, though not all will be used in neutral cases to complete a Delaunay triangulation. Also note that, in the non-neutral case, every Voronoi point is the common intersection of exactly three edges of the Voronoi diagram [68].

The following result for characterizing a Delaunay edge will be useful later when designing algorithms for constructing Delaunay triangulations. We assume here that neutral cases do not exist.

Theorem 3.1. *Any edge e_{ij} between two points p_i and p_j of a point set P is a Delaunay edge if and only if there exists a circle passing through p_i and p_j that does not contain any points of P in its interior.*

Proof. Any circle passing through p_i and p_j have its center v on the perpendicular bisector of a line between p_i and p_j, and v is equidistant to p_i and p_j. If the circle contains no point from P in its interior, the center v is closer to p_i and to p_j than to any other point from P. From the definition of a Voronoi region, it then follows that v lies on a Voronoi edge common to $V(p_i)$ and $V(p_j)$. Then p_i and p_j are Voronoi neighbors and consequently e_{ij} must be a Delaunay edge.

The converse now readily follows. If e_{ij} is a Delaunay edge between p_i and p_j, then p_i and p_j are Voronoi neighbors. Thus $V(p_i)$ and $V(p_j)$ share a Voronoi edge, and any circle with its center on the Voronoi edge common to $V(p_i)$ and $V(p_j)$ and passing through p_i and p_j does not contain any point from P in its interior.   □

If p_j is the closest point to p_i, then the line segment between p_i and p_j always defines a Delaunay edge. The other Delaunay edges depend on the distribution of P in the plane.

Let us explore some more of the important duality between the Voronoi diagram and the Delaunay triangulation. (Consult Figures 3.5 and 3.7). In the non-neutral case, a Voronoi point is equidistant from exactly three Voronoi neighbors in P. These three Voronoi neighbors define a Delaunay triangle. Therefore a Voronoi point is the center of the circumcircle of a Delaunay triangle. Thus, each Voronoi point can be associated with a triangle in the Delaunay triangulation. In a neutral case, a Voronoi point is equidistant from four or more points in P and it can be associated with two or more triangles.

Further, a Delaunay edge e_{ij} between two points p_i and p_j is always perpendicular to a Voronoi edge, although the two edges do not necessarily intersect. This Voronoi edge is the common edge shared by the Voronoi regions $V(p_i)$ and $V(p_j)$. Whether e_{ij} intersects its corresponding Voronoi edge depends on the location of p_i and p_j's Voronoi neighbors.

We summarize the duality between Voronoi diagrams and Delaunay triangulations as follows:

- *a Voronoi point corresponds to a Delaunay triangle,*
- *a Voronoi edge corresponds to a Delaunay edge,* and
- *a Voronoi polygon corresponds to a node in a Delaunay triangulation.*

We use these observations frequently in later sections when we develop the theory of Delaunay triangulations and algorithms for their construction. They are also used to compute a Voronoi diagram from a given Delaunay triangulation. Since the Delaunay triangulation is simpler to compute, the Voronoi diagram is usually derived from its dual triangulation. (See exercises 1, 2 and 3.)

3.5 The Circle Criterion

Figure 3.7 shows a Delaunay triangulation together with the Voronoi diagram and three of the circumcircles. The centers of the circles, which coincide with Voronoi points, are marked with small circles.

The circumcircles of the triangles in a Delaunay triangulation, have the following important property.

Lemma 3.1. *The circumcircle of a triangle in a Delaunay triangulation of a set of points P contains no point from P in its interior.*

Proof. We prove the lemma by contradiction. Consider the three points p_1, p_2 and p_3 in Figure 3.8 forming a triangle in a Delaunay triangulation, and the circimcircle $C(t_{1,2,3})$ of that triangle. The center v of $C(t_{1,2,3})$ is a vertex of the Voronoi diagram of P. Therefore, the Voronoi point v is common to the Voronoi regions $V(p_1)$, $V(p_2)$ and $V(p_3)$. If a point p_4 from P were inside $C(t_{1,2,3})$ as indicated in the figure, then v would be closer to p_4 than to any of p_1, p_2 and p_3. But then v would be inside $V(p_4)$ and not a Voronoi point of any of $V(p_1)$, $V(p_2)$ and $V(p_3)$, which is a contradiction. $\square$

This property is called the *circle criterion* and it leads to yet another definition of a Delaunay triangulation.

Definition 3.3 (Delaunay triangulation, circle criterion). *A Delaunay triangulation Δ of a set of points P in the plane is a triangulation where the interior of the circumcircle of any triangle in Δ contains no point from P.*

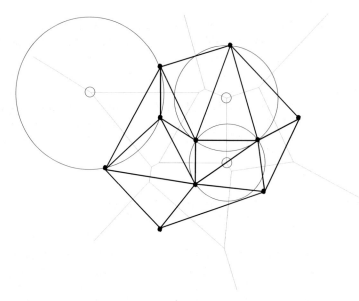

Fig. 3.7. A Voronoi diagram and its straight-line dual, the Delaunay triangulation, with some of the circumcircles.

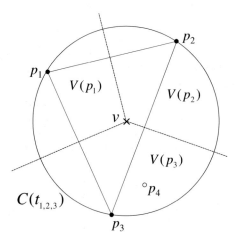

Fig. 3.8. The circimcircle $C(t_{1,2,3})$ of a triangle in a Delaunay triangulation and the Voronoi regions of p_1, p_2 and p_3. A point p_4 inside $C(t_{1,2,3})$ is not in agreement with the definition of a Voronoi diagram.

This definition is more geometric and possibly, from a practical point of view, more useful than the definition in Section 3.1. We will use the definition as a rule for constructing a Delaunay triangulation later when studying triangulation algorithms. Note that the circle criterion has the same neutral case as the MaxMin angle criterion discussed in Section 3.1. If we triangulate four distinct points on a common circle C, then there are two possible triangulations $\Delta = \{t_1, t_2\}$ and $\Delta' = \{t'_1, t'_2\}$ where the circumcircles of the triangles t_1, t_2, t'_1 and t'_2 all coincide with C and none of the points are in the interior of C. Thus the circle criterion holds for both triangulations.

3.6 Equivalence of the Delaunay Criteria for Strictly Convex Quadrilaterals

We have already proclaimed that the three different definitions of a Delaunay triangulation given earlier are equivalent. This will be proved later in this chapter. In this section, we consider triangulations consisting of two triangles only and prove that the definitions are equivalent in this simple case. The results will be used when proving equivalence in the general case and when developing algorithms for Delaunay triangulation.

Consider a *strictly convex* quadrilateral Q consisting of four points p_1, p_2, p_3 and p_4, see Figure 3.9. A quadrilateral is called strictly convex if each of its four interior angles are less than $180°$. There are two possible triangulations of Q depending on whether we connect p_1 and p_3 to form an edge, or connect p_2 and p_4 to form an edge. In the preceding sections we showed that the MaxMin angle criterion and the circle criterion have the same neutral case and thus the choice of the triangulation $\Delta = \{t_{1,2,4}, t_{2,3,4}\}$ or $\Delta' = \{t_{1,2,3}, t_{1,3,4}\}$ is arbitrary according to both criteria if the four points are cocircular. Below we show that the two criteria are also equivalent in the non-neutral case.

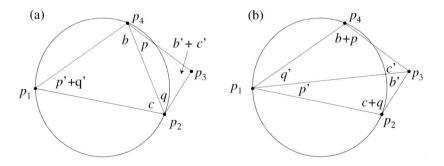

Fig. 3.9. Illustration for the proof of Lemma 3.2.

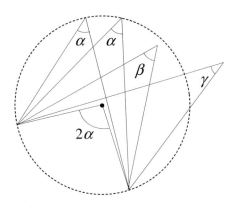

Fig. 3.10. Inscribed angles α on the same arc are equal, and $\gamma < \alpha < \beta$.

First we state some results from elementary geometry. Consider the angles in Figure 3.10. An inscribed angle of a circle is equal to half the central angle with the same arc as the inscribed angle. It then follows that all inscribed angles on the same arc are equal as indicated by the two angles with angular measure α. Further, the angles with measures β and γ in the figure must be greater than α and less than α respectively. We leave the formal proofs of these claims as exercises.

Lemma 3.2 ([81]). *Assume we have a strictly convex quadrilateral Q defined by a sequence of four points p_1, p_2, p_3 and p_4 that are not cocircular. The MaxMin angle criterion selects the edge (p_2, p_4) as the diagonal of Q if and only if p_3 lies outside the circumcircle of the triangle $t_{1,2,4}$, and it selects (p_1, p_3) as a diagonal if and only if p_3 lies inside the circumcircle.*

Proof. Consider the case where p_3 is outside the circumcircle as in Figure 3.9. Let $\Delta^{e_{2,4}}$ denote the triangulation on the left with the diagonal $e_{2,4}$ and let $\Delta^{e_{1,3}}$ denote the triangulation on the right with the diagonal $e_{1,3}$. Since p_3 is outside the circumcircle of $t_{1,2,4}$, the smallest interior angle in $\Delta^{e_{1,3}}$ must be one of b', c', p' and q'. Each of these primed angles has an unprimed counterpart in $\Delta^{e_{2,4}}$. It follows from elementary geometry that the angular measure of each primed angle is smaller than its unprimed counterpart. Further, the angles $p' + q'$ and $b' + c'$ in $\Delta^{e_{2,4}}$ are greater than the angles p' and b' respectively in $\Delta^{e_{1,3}}$. Then the triangulation of Q which results from choosing the edge $e_{2,4}$ must have a minimum interior angle which is strictly greater than the minimum interior angle of the triangulation constructed by choosing the edge $e_{1,3}$. Thus $e_{2,4}$ is preferred by the MaxMin angle criterion.

If p_3 lies inside the circumcircle, the strict inequalities of the angular measures are all reversed, and therefore the edge $e_{1,3}$ is preferred by the MaxMin angle criterion. $\square$

Note also the symmetry that p_3 is outside the circumcircle of $t_{1,2,4}$ if and only if p_1 is outside the circumcircle of $t_{2,3,4}$. The two triangles need not form a convex quadrilateral (Exercise 5).

In Section 3.4 we constructed the Delaunay triangulation from the Voronoi diagram by connecting Voronoi neighbors to form edges of the triangulation. This suggests that a third criterion, the *Voronoi region criterion* [50], can be applied for selecting a diagonal of the quadrilateral to make a Delaunay triangulation from Q. The following result links the Voronoi region criterion to the circle criterion. Since the circle criterion was shown to be equivalent to the MaxMin angle criterion, we find that the three Delaunay critera are equivalent for strictly convex quadrilaterals.

Lemma 3.3. *Suppose a strictly convex quadrilateral Q consisting of four points p_1, p_2, p_3 and p_4 that are not cocircular is given. Then p_2 and p_4 are strong Voronoi neighbors if and only if p_3 is strictly outside the circumcircle of the triangle $t_{1,2,4}$, and p_1 and p_3 are strong Voronoi neighbors if and only if p_3 is strictly inside the circumcircle.*

Proof. Consider first the Voronoi diagram of p_1, p_2 and p_4 shown in Figure 3.11(a). Then p_2 and p_4 are (trivially) strong Voronoi neighbors because the Voronoi regions $V(p_2)$ and $V(p_4)$ share a common Voronoi edge $e_{v,\infty}$. The Voronoi edge $e_{v,\infty}$ has an endpoint in the center v of the circumcircle $C(t_{1,2,4})$ and extends to infinity on the opposite side of the triangle edge $e_{2,4}$ from p_1. Assume that p_3 is added outside $C(t_{1,2,4})$ such that p_1, p_2, p_3 and p_4 form a strictly convex quadrilateral. Then $d(v, p_3) > d(v, p_i)$, $i = 1, 2, 4$, and thus v cannot be contained in the Voronoi region of p_3. But a segment of the Voronoi edge $e_{v,\infty}$ remains after p_3 has been added and thus p_2 and p_4 remain as strong Voronoi neighbors, see Figure 3.11(b).

To prove the converse, we observe that if p_2 and p_4 are strong Voronoi neighbors, then $e_{2,4}$ is a Delaunay edge and $\Delta = \{t_{1,2,4}, t_{2,3,4}\}$ must be the Delaunay

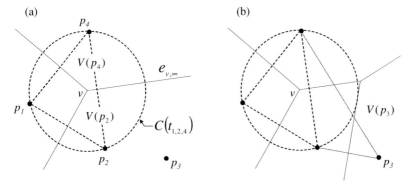

Fig. 3.11. Illustration for the proof of Lemma 3.3.

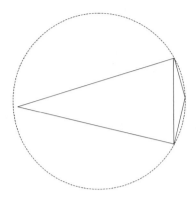

Fig. 3.12. Example showing that choosing the diagonal with the minimum length violates the circle criterion.

triangulation of the points of Q. Then, by Lemma 3.1, p_3 cannot be inside the circumcircle of $t_{1,2,4}$. $\qquad\square$

In the neutral case, the endpoints of the two diagonals of Q are pairwise weak Voronoi neighbors, so the choice of $e_{2,4}$ or $e_{1,3}$ as an edge of the Delaunay triangulation is arbitrary. Thus the Voronoi region criterion has the same neutral case as the other criteria.

Note that a fourth criterion, that of choosing the diagonal of a quadrilateral with the minimum length, has been suggested, and even claimed to be a Delaunay criterion. The example in Figure 3.12 is a counterexample to this claim. Here the longer diagonal must be chosen, as the other diagonal clearly violates the circle criterion.

3.7 Computing the Circumcircle Test

The algorithmic approach to determining which diagonal in a strictly convex quadrilateral should be chosen for the Delaunay criteria would typically be to examine the interior angles of the quadrilateral. If α and β are the two interior angles opposite the existing diagonal as shown in Figure 3.13, then the other diagonal must be chosen if and only if $\alpha + \beta > \pi$. Since $\alpha + \beta < 2\pi$, this is equivalent to $\sin(\alpha + \beta) < 0$ which expands to

$$\sin \alpha \cos \beta + \cos \alpha \sin \beta < 0. \tag{3.3}$$

This test, like other equivalent tests, is commonly known as the *circumcircle test* or the *incircle test*. If $\alpha + \beta = \pi$ there is a neutral case and the choice of diagonal is arbitrary. Note that this approach also applies for quadrilaterals that are not convex such that testing for convexity is not necessary (Exercise 7).

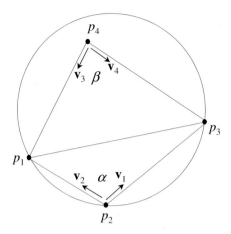

Fig. 3.13. Illustration for circumcircle test based on interior angles.

Recall the definition of scalar product and cross product between vectors. The cross product of three-dimensional vectors $\mathbf{a} = (a_1, a_2, a_3)$ and $\mathbf{b} = (b_1, b_2, b_3)$ gives a vector in 3D space and is defined by

$$\mathbf{a} \times \mathbf{b} = (a_2 b_3 - a_3 b_2, a_3 b_1 - a_1 b_3, a_1 b_2 - a_2 b_1).$$

The scalar product, which gives a scalar, is defined by

$$\mathbf{a} \cdot \mathbf{b} = a_1 b_1 + a_2 b_2 + a_3 b_3.$$

Suppose that the vectors $\mathbf{v}_i$, $i = 1, 2, 3, 4$, in Figure 3.13 are unit vectors. For common notation we assume that the vectors and the four points of the quadrilateral are defined in 3D space. For example, $p_1 = (x_1, y_1, 0)$ and $\mathbf{v}_1 = (x_3 - x_2, y_3 - y_2, 0) / \|p_3 - p_2\|_2$, where $\|\cdot\|_2$ denotes the Euclidean norm. Let $\mathbf{e}_3$ be the unit vector $(0, 0, 1)$. Then, sine and cosine of α and β for the inequality in (3.3) can be computed by cross products and scalar products,

$$\sin \alpha = (\mathbf{v}_1 \times \mathbf{v}_2) \cdot \mathbf{e}_3 = \frac{(x_3 - x_2)(y_1 - y_2) - (x_1 - x_2)(y_3 - y_2)}{\|p_3 - p_2\|_2 \, \|p_1 - p_2\|_2}$$

$$\sin \beta = (\mathbf{v}_3 \times \mathbf{v}_4) \cdot \mathbf{e}_3 = \frac{(x_1 - x_4)(y_3 - y_4) - (x_3 - x_4)(y_1 - y_4)}{\|p_1 - p_4\|_2 \, \|p_3 - p_4\|_2}$$

$$\cos \alpha = \mathbf{v}_1 \cdot \mathbf{v}_2 = \frac{(x_3 - x_2)(x_1 - x_2) + (y_3 - y_2)(y_1 - y_2)}{\|p_3 - p_2\|_2 \, \|p_1 - p_2\|_2}$$

$$\cos \beta = \mathbf{v}_3 \cdot \mathbf{v}_4 = \frac{(x_1 - x_4)(x_3 - x_4) + (y_1 - y_4)(y_3 - y_4)}{\|p_1 - p_4\|_2 \, \|p_3 - p_4\|_2}.$$

The denominators in these expressions do not need to be computed because the two terms in the relation (3.3) have a common denominator.

Great care should be taken to ensure numerical stability of the circumcircle test. When there are almost neutral cases, or when points of a quadrilateral are almost coincident or almost collinear, round-off errors may lead to incorrect results. This may again lead to cycling and infinite loops in algorithms. The test in (3.3) can be improved by using multiple and more robust tests, or by using exact or "almost exact" arithmetic [78]. Algorithm 3.1, which determines if the diagonal edge in Figure 3.13 should be swapped, is taken from Cline & Renka [17], where it is claimed that the algorithm ensures numerical stability. In Step 5, zero could be replaced by a small negative number to avoid swapping in nearly neutral cases.

Algorithm 3.1 Circumcircle test

---
1. **if** $(\cos\alpha < 0$ and $\cos\beta < 0)$
2.   **return** FALSE   // swap the edge
3. **if** $(\cos\alpha > 0$ and $\cos\beta > 0)$
4.   **return** TRUE
5. **if** $(\cos\alpha\sin\beta + \sin\alpha\cos\beta < 0)$
6.   **return** FALSE   // swap the edge
7. **else**
8.   **return** TRUE

The circumcircle test in (3.3) has an equivalent form based directly on the coordinates of the points $p_i = (x_i, y_i, 0)$, $i = 1, 2, 3, 4$, of the quadrilateral in Figure 3.13. The test can be implemented by evaluating the sign of the determinant,

$$\mathcal{D} = \begin{vmatrix} x_1 & y_1 & x_1^2 + y_1^2 & 1 \\ x_2 & y_2 & x_2^2 + y_2^2 & 1 \\ x_3 & y_3 & x_3^2 + y_3^2 & 1 \\ x_4 & y_4 & x_4^2 + y_4^2 & 1 \end{vmatrix}. \tag{3.4}$$

If $\mathcal{D} > 0$, this is equivalent to the inequality in (3.3) and the situation in Figure 3.13. Thus, the other diagonal must be chosen when $\mathcal{D} > 0$ to fulfil the Delaunay criteria.

3.8 The Local Optimization Procedure (LOP)

Lawson [50] suggests a simple *local optimization procedure (LOP)* for constructing a new triangulation from an existing triangulation. The procedure operates on the edges of the triangulation and considers edge-swaps based on the criteria discussed in the preceding sections. More specifically, let e_i denote an interior edge of a triangulation and let Q be the quadrilateral formed by the two triangles having e_i as a common edge. If Q is not strictly convex, then e_i cannot be swapped, and in the neutral case when the four points of Q lie on a common circle, the decision is not to swap e_i. Otherwise, apply any of the equivalent criteria discussed in the previous section and swap e_i if this is preferred by the criterion.

Definition 3.4. *An edge in a triangulation will be called locally optimal if the decision is not to swap it when applying the LOP.*

This also implies that an edge e_i becomes locally optimal just after it has been swapped. The edges on the boundary of the triangulation are locally optimal by default. In the following, let $I(\Delta)$ denote the indicator vector of a triangulation Δ as defined in Section 3.1.

Lemma 3.4 ([50]). *Assume that a triangulation Δ is given and let e_i be an interior edge of Δ. Suppose application of the LOP to e_i results in a swap, replacing e_i by a new edge e_i', and thus replacing Δ by a new triangulation Δ'. Then $I(\Delta') > I(\Delta)$, that is, Δ' strictly follows Δ in the linear ordering.*

Proof. Let α_i and α_j be the two entries of $I(\Delta)$ representing the smallest angles of each of the two triangles sharing the edge e_i, and assume that $i < j$ and thus $\alpha_i \leq \alpha_j$; $I(\Delta) = (\alpha_1, \ldots, \alpha_i, \ldots, \alpha_j, \ldots, \alpha_{|T|})$. Since a swap was made when applying the LOP on e_i, the smallest interior angle of the two new triangles sharing the edge e_i' must be strictly greater than α_i according to the MaxMin angle criterion. It follows that the indicator vector $I(\Delta')$ of the new triangulation Δ' is lexicographically larger than $I(\Delta)$ and Δ' strictly follows Δ in the linear ordering. □

Applying the LOP repeatedly to the edges of a triangulation, the theorem states that the indicator vector $I(\Delta)$ becomes lexicographically larger each time an edge-swap occurs. This implies that cycling, where the same triangulation is reached several times, cannot occur. Note however that an edge e_i can be swapped more than once in this process. This may happen if one of the edges of the quadrilateral with e_i as a diagonal is swapped after e_i has been swapped, in which case e_i might not be locally optimal any more.

Since the number of possible triangulations of a finite point set is finite, the LOP converges after a finite number of edge-swaps. The result is a triangulation where all edges are locally optimal.

Definition 3.5. *A triangulation whose edges are all locally optimal (after applying the LOP) is said to be a locally optimal triangulation.*

In general, for arbitrary swapping criteria such as the MinMax angle criterion, the locally optimal triangulation obtained by the LOP (Algorithm 3.2) depends on the specific order in which edges are swapped. Thus the LOP does not necessarily yield a triangulation whose indicator vector is lexicographically

---

Algorithm 3.2 LOP, Local Optimization Procedure

---

1. Make an arbitrary legal triangulation Δ of a point set P.
2. **if** Δ is locally optimal,
   stop.
3. Let e_i be an interior edge of Δ which is not locally optimal.
4. Swap e_i to e_i', thus transforming Δ to Δ'.
5. Let $\Delta := \Delta'$.
6. **goto** 2.

---

maximum since it may terminate at local extrema. However, for the Delaunay criteria discussed in the previous section, we will show later that the LOP yields a triangulation whose indicator vector is always lexicographically maximum, and thus by Definition 3.1, it is also a Delaunay triangulation.

3.9 Global Properties of the Delaunay Triangulation

In the preceding sections, we established three different definitions of a Delaunay triangulation, each with an accompanying swapping criterion, and studied local properties of Delaunay triangles and Delaunay edges. In this section, we establish the theoretical basis which provides necessary results for designing algorithms for Delaunay triangulation. The three definitions of a Delaunay triangulation given earlier will be linked together in a unified theory. We will see that the result from applying the LOP has important global consequences. More specifically, applying the LOP repeatedly results in a triangulation that is *globally optimal*, that is, its indicator vector is lexicographically maximum. Moreover, we prove that the globally optimal triangulation is in fact a Delaunay triangulation as defined by the straight-line dual of the Voronoi diagram in Section 3.4.

To avoid lengthy details and explanations in the proofs of the theorems, we assume that no more than three points in a point set P are cocircular and that the points in P are not all collinear. Then, with the constructive definition of a Delaunay triangulation from the Voronoi diagram in Section 3.4, we also claim that the Delaunay triangulation is unique. That is, there is only one possible Delaunay triangulation of a finite point set P with no more than three cocircular points.

First we show that if all edges in a triangulation are locally optimal, that is, if the triangulation is locally optimal, then the triangulation is Delaunay as defined in Definition 3.3.

Theorem 3.2 ([50]). *All interior edges of a triangulation Δ of a point set P are locally optimal if and only if no point from P is interior to any circumcircle of a triangle in Δ.*

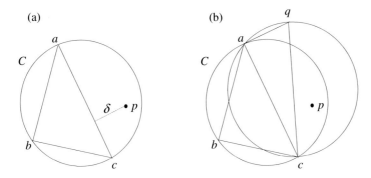

Fig. 3.14. Illustration for the proof of Theorem 3.2.

Proof. If no point of P is interior to any circumcircle of P, then application of the LOP to any edge of Δ will not result in a swap. Thus all edges of Δ are locally optimal.

To prove the converse, assume all edges of Δ are locally optimal. Suppose for the sake of contradiction there is a triangle $t_{a,b,c}$ with circumcircle C that has a point p from P in its interior. The point p cannot belong to any of the quadrilaterals formed by $t_{a,b,c}$ and another triangle as this would contradict the hypothesis that all edges are locally optimal. Let δ be the perpendicular distance from p to the nearest edge of $t_{a,b,c}$, say $e_{a,c}$, as shown in Figure 3.14(a). Among all triangles of Δ whose circumcircle contains p as an interior point, assume that none has an edge at a distance less than δ from p. Since p is on the opposite side of $e_{a,c}$ from b, $e_{a,c}$ is not a boundary edge. Thus there is another triangle $t_{a,c,q}$ sharing the edge $e_{a,c}$ with $t_{a,b,c}$ as shown in Figure 3.14(b). The vertex q cannot be interior to the circle C as this would contradict the hypothesis that $e_{a,c}$ is locally optimal (and p cannot be inside $t_{a,c,q}$). Assume that $e_{c,q}$ is the closest edge of $t_{a,c,q}$ to p as in the figure. The distance from $e_{c,q}$ to p is now clearly less than δ. Since the circumcircle of $t_{a,c,q}$ must also contain p in its interior, we have reached a contradiction to the hypothesis that $t_{a,b,c}$ is the triangle with an edge at the smallest distance from p and with a circumcircle that contains p. □

Recall that when applying the LOP repeatedly to the edges of a triangulation, the LOP converges and all the edges of the resulting triangulation are locally optimal. Thus, from the theorem above and Definition 3.3 we can now conclude that the LOP converges to a (locally optimal) triangulation that is Delaunay according to Definition 3.3.

In the following, we establish the relationship between a locally optimal triangulation of a point set P and a Delaunay triangulation of P as defined by the straight-line dual of the Voronoi diagram. First we make a stepping stone towards the final results.

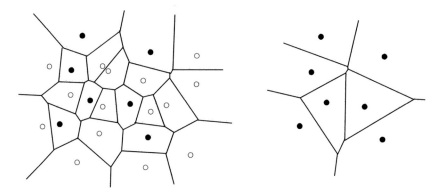

Fig. 3.15. Voronoi diagram of a point set P on the left, and Voronoi diagram of a subset $\widetilde{P} \subset P$ on the right.

Consider the two Voronoi diagrams in Figure 3.15. On the Left, the Voronoi diagram of a point set P is shown. Some of the points are represented as filled bullets, which imply that they belong to a subset $\widetilde{P} \subset P$. On the right, the Voronoi diagram of only the subset $\widetilde{P}$ is shown. The important observation here is that Voronoi neighbor relationships between points in the subset $\widetilde{P}$ in the left diagram are preserved in the right diagram when all points from P that do not belong to $\widetilde{P}$ have been removed. We state this formally by a lemma as an aid for proofs of theorems later in this chapter and in other chapters.

Lemma 3.5 ([50]). *Let P be a set of points in the plane and let $\widetilde{P}$ be a subset of P. Two points in $\widetilde{P}$ that are strong Voronoi neighbors in P will also be strong Voronoi neighbors in $\widetilde{P}$.*

Proof. Consider the effect of removing one point, say p, from P that is not in $\widetilde{P}$, leaving a point set P'. The only effect of this process is that the Voronoi region for p is absorbed into the Voronoi regions of neighboring Voronoi regions. This change takes place such that no Voronoi edge between the remaining Voronoi regions is shortened. Thus, two points in P' that are strong Voronoi neighbors in P will also be strong Voronoi neighbors in P'. The proof follows by considering the process of removing one point at a time until the subset $\widetilde{P}$ is left. □

Let us also consider the dual Delaunay triangulations in view of this result. Let $\Delta(P)$ be the Delaunay triangulation of P and let $\Delta(\widetilde{P})$ be the Delaunay triangulation of the subset $\widetilde{P}$ of P. Then any edge of $\Delta(P)$ whose endpoints are both in $\widetilde{P}$ is also an edge in $\Delta(\widetilde{P})$.

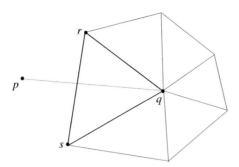

Fig. 3.16. Illustration for the proof of Theorem 3.3.

Theorem 3.3 ([50]). *All interior edges of a triangulation Δ of a point set P are locally optimal if and only if Δ is a Delaunay triangulation as defined by the straight-line dual of the Voronoi diagram.*

Proof. If Δ is a Delaunay triangulation, then by Lemma 3.1 no point of P is interior to any circumcircle of a triangle in Δ. By Theorem 3.2, all interior edges are then locally optimal.

   To prove the converse, assume all edges of Δ are locally optimal. Suppose that the theorem is false, that is, suppose that Δ is not a Delaunay triangulation. Then there must be two strong Voronoi neighbors in P, say p and q that are not connected by an edge in Δ. Consider the set of triangles that have q as a common vertex, see Figure 3.16. The line segment $\overline{pq}$ must clearly intersect one of the triangle edges opposite q. This is also the case if q is at the boundary of Δ since the boundary is convex. Moreover, since p and q are Voronoi neighbors in P, there can be no other point of P on the line segment $\overline{pq}$. Thus $\overline{pq}$ must intersect the interior of one of the triangle edges opposite q, say the edge $e_{r,s}$ as in the figure. By hypothesis, all interior edges of Δ are locally optimal. Then, by Theorem 3.2, the point p cannot be interior to the circumcircle of the triangle $t_{q,r,s}$. By the equivalence of the circle criterion and the Voronoi region criterion for strictly convex quadrilaterals, p and q cannot be strong Voronoi neighbors relative to the subset $\{p, s, q, r\}$ of P. However, by Lemma 3.5, p and q must be strong Voronoi neighbors in the subset $\{p, s, q, r\}$ since they are strong Voronoi neighbors in P, and we have reached a contradiction. $\square$

Corollary 3.1. *A triangulation Δ of a set of points P is a Delaunay triangulation if and only if no point of P is interior to any circumcircle of a triangle in Δ.*

   The last result, which follows directly from the two preceding theorems, is probably the most frequently used characterization of a Delaunay triangulation. By the above theorem we have shown that applying the local optimization procedure by Lawson repeatedly on strictly convex quadrilaterals, will eventually result in a triangulation that is a Delaunay triangulation. This also implies another important property of the edge-swapping process that we have mentioned earlier.

Theorem 3.4. *Given an arbitrary triangulation Δ of a point set P, where the boundary of Δ is convex. Then any other triangulation Δ' of P, with the same boundary as Δ, can be reached by a sequence of edge-swaps starting from Δ.*

Proof. From Theorem 3.3, applying the LOP repeatedly to the edges of Δ and Δ' gives the same result which is the Delaunay triangulation of P. This follows from the uniqueness of a Delaunay triangulation when we assume that no four points are cocircular. Since a single edge-swap clearly is reversible, a sequence of edge-swaps is also a reversible process. Then Δ' can be reached by a sequence of edge-swaps starting from the Delaunay triangulation of P which again can be reached by a sequence of edge-swaps starting from Δ'.   □

Again, we do not explain the details in the case where four or more points in P are cocircular; a complete proof can be found in [49]. It can also be shown that the theorem holds in the case of a non-convex boundary, see [23] which also gives a detailed algorithm for transforming Δ into Δ'. We will return to this in Chapter 6 when we define constrained edges between nodes in a triangulation.

The final result of this section shows that the Delaunay triangulation is also globally optimal, that is, a locally optimal triangulation is also globally optimal when using one of the equivalent Delaunay criteria and assuming that no four points are cocircular.

Theorem 3.5. *A triangulation Δ of a set of points P is a Delaunay triangulation as defined by the straight-line dual of the Voronoi diagram, if and only if its indicator vector $I(\Delta)$ is lexicographically maximum. That is, no triangulation follows it in the linear ordering.*

Proof. If the indicator vector of Δ is lexicographically maximum, then all the edges of Δ must be locally optimal. By Theorem 3.3, Δ must then be a Delaunay triangulation.

To prove the converse, assume Δ is a Delaunay triangulation. Suppose for the sake of contradiction that $I(\Delta)$ is not lexicographically maximum in the linear ordering. Since by hypothesis Δ is a Delaunay triangulation, all interior edges must be locally optimal by Theorem 3.3. If $I(\Delta)$ is not lexicographically maximum, there exists another triangulation Δ' of P such that $I(\Delta') > I(\Delta)$. If the edges of Δ' are not all locally optimal, we can apply the LOP repeatedly to Δ' and thus obtain a triangulation Δ^* where all edges are locally optimal and such that $I(\Delta^*) > I(\Delta)$. By Theorem 3.3, Δ^* must also be a Delaunay triangulation. However, since the Delaunay triangulation is unique, $I(\Delta^*) = I(\Delta)$, and we have reached a contradiction.   □

3.10 Exercises

1. Explain how to "draw" a Voronoi diagram overlaid on an existing Delaunay triangulation. Is there any ambiguity or is the Voronoi diagram unique by this construction?
2. Describe a strategy (e.g. by pseudo-code) for constructing a Voronoi diagram from a given Delaunay triangulation. Avoid computing line-line intersections.
3. Assume that a Delaunay triangulation $\Delta(P)$ is computed and represented in an appropriate data structure. Add details to the strategy in Exercise 2 to obtain an algorithmic design for finding the Voronoi diagram of P based on $\Delta(P)$. Let the implementation contain functionality for finding the Voronoi polygon of a point in P and a function for printing the Voronoi diagram (e.g. to a file). Avoid printing Voronoi edges twice.
4. The circumcenter of a triangle is the center of the triangle's circumscribed circle. Assume that all triangles in a triangulation have the property that each circumcenter falls inside its own triangle. Show that the triangulation is Delaunay.
5. Let t_1 and t_2 be two adjacent triangles in a Delaunay triangulation sharing an edge e_i. Further, let p_1 and p_2 be the vertices of t_1 and t_2 respectively, on the opposite side of e_i. Show that p_2 is outside the circumcircle of t_1 if and only if p_1 is outside the circumcircle of t_2; (t_1 and t_2 need not form a convex quadrilateral).
6. Let p_i and p_j be two points in a point set P, and let p_j be the closest point to p_i. Show that p_i, p_j form a Delaunay edge.
7. Let α and β be the two interior angles of a quadrilateral opposite an existing diagonal as in Figure 3.13. Verify that the other diagonal must be chosen to fulfil the Delaunay criteria if and only if $\sin\alpha\cos\beta + \cos\alpha\sin\beta < 0$, and that this also applies for quadrilaterals that are not strictly convex.
8. Show that inscribed angles on the same arc of a circle are equal, see Figure 3.10.
9. Verify that the equivalent form of the circumcircle test using the determinant in (3.4) holds.

4

Algorithms for Delaunay Triangulation

Several algorithms have been developed for Delaunay triangulation based on the definitions and the theory of the previous chapter. The popularity of the Delaunay triangulation is twofold. It yields "good shaped" triangles (in the plane) and the theory, mainly based on its dual, the Voronoi diagram, is well established. Other types of triangulation, such as triangulations that are optimal in the sense of the MinMax angle criterion, are difficult to compute in reasonable time from a large number of points. In fact, the Delaunay swapping criteria, which were shown to be equivalent in Section 3.6, are the only known criteria that can be used in Lawson's local optimization procedure (LOP) to guarantee a globally optimal triangulation.

In this chapter different approaches for computing Delaunay triangulations are presented. Some of the algorithms are outlined briefly without details on underlying data structures. The theory developed in the previous chapter is frequently referred to, and more theory is established to ascertain that the algorithms produce triangulations that are Delaunay, as defined previously.

4.1 A Simple Algorithm Based on Previous Results

In Section 1.4 we introduced simple algorithms for constructing a triangulation with a given boundary from a set of points in the plane. Further, in Section 3.8 we introduced a local optimization procedure (LOP) for updating a triangulation according to the Delaunay criteria such that the triangulation obtained on termination of the LOP was Delaunay. Assume that the given boundary is the convex hull of a set of points P in the plane which we denote $conv(P)$. There are many algorithms for computing $conv(P)$. Its construction requires $O(N \log N)$ time, see for example [68]. We can now use these building blocks to obtain a simple Delaunay triangulation algorithm of P (Algorithm 4.1).

Since the final triangulation Δ^* constructed with this algorithm has all its edges locally optimal, it is a Delaunay triangulation by Theorem 3.3. Although

---

Algorithm 4.1 Simple Delaunay triangulation

1. Compute $conv(P)$.
2. Apply Algorithm 1.1 to the vertices of $conv(P)$ to find an initial triangulation Δ'.
3. Apply Algorithm 1.2 to insert points of P, that are interior to $conv(P)$, into Δ'. This gives a new triangulation Δ'' after all points have been inserted.
4. Apply the LOP repeatedly on the edges of Δ'' until no edge-swap occurs, and thus obtain a triangulation Δ^* which has all of its edges locally optimal.

---

simple and fairly easy to implement if a proper data structure has been established, this algorithm is not optimal regarding running time. Also, Step 3 of the algorithm may accumulate triangles that are degenerate or almost degenerate when a point is inserted near an existing edge. This may result in numerical instability when a new point is inserted in Step 3 or when the LOP is run in Step 4. The other algorithms outlined in this chapter will be more robust and have better average performance.

4.2 Radial Sweep

There are many algorithms in the literature which use a similar approach to the above for constructing a Delaunay triangulation. The algorithms differ on how they make the initial triangulation prior to the local optimization procedure in Step 4. The *Radial Sweep* algorithm replaces Step 1, 2 and 3 with procedures that also find the convex hull in the plane of a point set P.

Algorithm 4.2 Radial Sweep

1. Choose a point p near the centroid of P and connect p by radiating edges to all other points of P, Figure 4.1(a).
2. Sort and order the points $\{P \backslash p\}$ by orientation and distance from p and connect the ordered sequence by edges as in Figure 4.1(b). The result from this step is a triangulation with a star-shaped domain as seen from p. Triangles may be degenerate since points may have identical orientation relative to p.
3. Form a triangle for each triple of points (p_{i-1}, p_i, p_{i+1}) on the boundary of the triangulation. If the edge between p_{i-1} and p_{i+1} is outside the existing boundary, include the triangle in the triangulation and update the boundary. Repeat this step until no more triangles can be added. The resulting triangulation has a convex boundary and all points are included in the triangulation, Figure 4.1(c).
4. Apply the LOP repeatedly to the edges until no edge-swap occurs, to obtain a Delaunay triangulation, Figure 4.1(d).

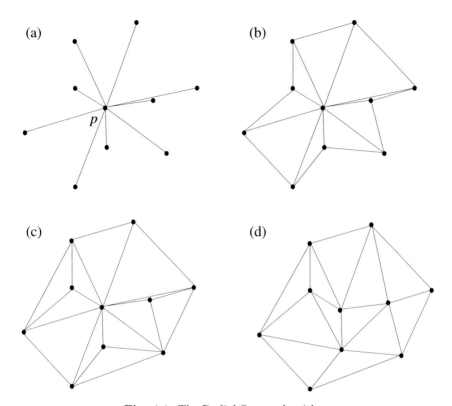

Fig. 4.1. The Radial Sweep algorithm.

Originally, as presented in [64], Step 4 of Algorithm 4.2 uses the minimum diagonal criterion when considering edge-swaps. As we concluded in Section 3.6, this criterion is not equivalent to the Delaunay criteria. However, in most cases it will make the same decision as the Delaunay swapping criteria when determining edge-swaps, so the result may be an "almost Delaunay triangulation". Also, comparing lengths of diagonals (edges) is faster than the calculations involved when using the Delaunay criteria, so it is worthwhile considering the minimum diagonal criterion when speed is important.

4.3 A Step-by-Step Approach for Making Delaunay Triangles

The step-by-step approach builds the Delaunay triangulation successively by constructing one triangle at a time (or one edge at a time), finally producing a Delaunay triangulation. Many algorithms have been based on this principle, the first was probably by McLain [59]. The algorithm starts from an edge e_b,

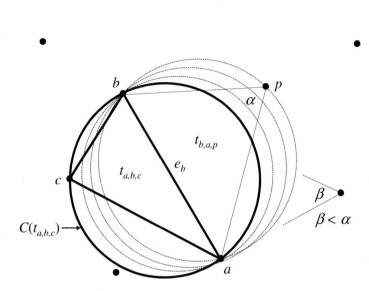

Fig. 4.2. Circle growing from the base line e_b to find a point p to form a new triangle with e_b.

a *base line*, formed by two Voronoi neighbors of a point set P such that e_b is a Delaunay edge with respect to P. (See Figure 4.2.) For example, e_b can be a boundary edge of the convex hull of P, or the endpoints p_i and p_j of e_b can be such that p_j is the nearest neighbor of p_i, or vice versa. (Exercise 2.) Next, e_b is connected to a point of P, say p, to form a new triangle. This is done successively by choosing a new base line e_b on the boundary of the existing triangulation in each step.

The main feature of this algorithm is that each new edge and each new triangle that is formed by connecting e_b to p will be edges and triangles of the final Delaunay triangulation $\Delta(P)$. Thus, p cannot be an arbitrary point of P, and all points of P that are not interior nodes of the current triangulation must be considered somehow in each step. In fact, by the uniqueness of the Delaunay triangulation, p is a unique point if four or more points of P are not cocircular. (We make the same assumption here.)

Before outlining the algorithm, we need a result that uniquely determines a Delaunay triangle. We already know from Lemma 3.1 that the circumcircle of a Delaunay triangle is *point-free*, that is, the circle does not contain any point from P in its interior. But does the converse also hold? Given three points a, b and c of a point set P such that the triangle $t_{a,b,c}$ formed by these three points is point-free, can we then guarantee that $t_{a,b,c}$ is a triangle of the

Delaunay triangulation of P? We extend Lemma 3.1 to a theorem that states that this holds and leave the proof to the reader.

Theorem 4.1. *A triangle $t_{a,b,c}$ with nodes a, b and c from a point set P is a Delaunay triangle if and only if the circumcircle of $t_{a,b,c}$ contains no point from P in its interior.*

Based on this theorem and the theory derived in the previous chapter, we can find the unique point p to form a new Delaunay triangle with a base line e_b by considering a *growing circle* from e_b.

Assume first that a Delaunay triangle $t_{a,b,c}$ has already been found as depicted in Figure 4.2. By Theorem 4.1 (and Lemma 3.1) the circumcircle C of $t_{a,b,c}$ is point-free. Let e_b, with endpoints a and b, be an edge of $t_{a,b,c}$ that is not shared by another triangle. Imagine that C grows on the opposite side of e_b from c while still interpolating a and b (see the dotted circles in the figure). The center of the growing circle moves along the perpendicular bisector of the line between a and b. If a point from P is never reached by the growing circle, then e_b must be an edge of the convex hull of P and another edge must be chosen as the base line e_b. Let p be the first point of P that is reached by the growing circle and connect p to e_b to form a new triangle $t_{b,a,p}$. Note that p can either be a point that is not yet included as a node in the current triangulation, or p can be a node of a boundary edge of the triangulation from the previous step. Since the growing circle never exceeds the circumcircle of $t_{a,b,c}$ on the opposite side of e_b from p, the growing circle is still point-free when p is reached. Moreover, the growing circle has now become the circumcircle of $t_{b,a,p}$. Therefore, by Theorem 4.1 the new triangle $t_{b,a,p}$ is a Delaunay triangle.

Repeatedly applying this process with e_b as a boundary edge of the current triangulation, a new Delaunay triangle is created each time unless e_b is an edge of the convex hull of P. The process terminates when no new edge e_b that is not an edge of the convex hull of P can be found. The result is a triangulation $\Delta(P)$ where all triangles are Delaunay and where the boundary of the triangulation is the convex hull of P. Thus, $\Delta(P)$ is a Delaunay triangulation of the point set P.

Since the algorithm must start with a single Delaunay edge, the first point-free circle can be centered at the midpoint of e_b if the endpoints of e_b are two nearest neighbors of P. If e_b is an edge of the convex hull of P, the first point-free circle can be infinitely large with its center on the perpendicular bisector of e_b outside the convex hull of P. The algorithmic approach for the growing circle would be to examine the angle α at a point p and spanned by the endpoints of e_b, see Figure 4.2. The point making the largest angle will be chosen to form a new triangle with e_b.

4.4 Incremental Algorithms

The Radial Sweep and step-by-step algorithms are *static* in the sense that all points of P must be known from the first step of the algorithms. Moreover, all points of P must be considered somehow in each step. The "simple" approach in Section 4.1 only needs to know the points forming the convex hull, $conv(P)$, of P before points from P are inserted into the interior of $conv(P)$. This algorithm can be modified slightly to apply the LOP after each point insertion to fulfill the Delaunay criteria in each step. Similar approaches have been used in many algorithms that we will call *incremental*[1] algorithms [34, 52, 86].

Incremental algorithms start with an initial triangulation that can be a single triangle, one or more large enclosing triangles, or a triangulation of $conv(P)$. Then points of P are inserted interior or exterior to the initial triangulation. If the initial triangulation is formed by enclosing triangles with nodes that are not in P, these points must be removed in a last step in such a way that the final triangulation is also Delaunay (with convex boundary).

The point insertion is characterized by the facts that,

- the remaining points of P that have not already been inserted, need not be considered, and
- after each point insertion, the triangulation is updated to be Delaunay.

In general, points can be inserted in any order, but many algorithms are based on sorting P before the triangulation process starts. Then a qualified selection of the insertion point can be made for optimizing running time. For example, points may be sorted lexicographically on x and y such that each new point is inserted outside the existing boundary, or such that each insertion point is expected to be close to the previous inserted point.

Incremental algorithms need efficient procedures for inserting points into a triangulation. When inserting a point p into an existing Delaunay triangulation Δ_N of a point set P with N points, there are conceptually two ways to update Δ_N to become a new triangulation Δ_{N+1} that is also Delaunay.

(a) The collection of triangles R^p of Δ_N that need to be modified by the insertion of p into Δ_N is determined, see Figure 4.3. All triangles inside R^p are removed. The points from P in R^p and the insertion point p are retriangulated such that the resulting triangulation Δ_{N+1} is Delaunay.

(b) The triangle t of Δ_N containing p is located, see Figure 4.4. A general algorithm for locating a triangle containing a given point is described in detail in Chapter 9. The nodes of t are connected to p to form three new edges in a triangulation Δ'_{N+1} which is not necessarily Delaunay. Then a swapping procedure is applied to the edges of Δ'_{N+1} such that all edges become locally optimal. The final triangulation Δ_{N+1} will then

---

[1] Some authors also call them *iterative* algorithms.

be a Delaunay triangulation by Theorem 3.3. If p lies outside Δ_N, the initial triangulation Δ'_{N+1} is obtained by connecting p to all nodes at the boundary of Δ_N that are visible from p.

In both cases it is important to know which part of Δ_N that needs to be modified to obtain Δ_{N+1}. When using the approach in (a), an exact limitation of R^p must be found, and a precise procedure for retriangulating R^p is needed to ensure that Δ_{N+1} is Delaunay. Using the method in (b), the work to be done by the LOP must be limited such that not all edges of the triangulation need to be tested for local optimality each time a point is inserted. We investigate this in detail in the following.

4.5 Inserting a Point into a Delaunay Triangulation

Fortunately, inserting a point p into a Delaunay triangulation appears to be a local process. In most cases only a limited region near p needs to be rearranged when inserting p. We call the region R^p in Δ_N that needs to be modified by the insertion of p, the *influence region* of point p in Δ_N. The external boundary of R^p formed by triangle edges and denoted Q^p, is called the *influence polygon* of point p in Δ_N [21], see Figure 4.3.

In the following, we show how to find an exact limitation of the influence region R^p and how to retriangulate R^p when a point is inserted. To avoid detailed explanations, we assume that no four points are cocircular and that not all points are collinear. The reader should also keep in mind the existence and the uniqueness of a Delaunay triangulation as discussed in Section 3.4. Also recall that a triangle is a Delaunay triangle if and only if there are no points interior to its circumcircle.

Lemma 4.1. *A triangle t in a Delaunay triangulation Δ_N will be modified when inserting a point p to obtain a Delaunay triangulation Δ_{N+1} if and only if the circumcircle of t contains p in its interior.*

Proof. The proof follows immediately from Theorem 4.1.   □

This shows that the influence region R^p is limited to the triangles of Δ_N whose circumcircles contain p in their interior. Figure 4.3(a) and (c) show influence regions when a point is inserted into the interior of an existing Delaunay triangulation and outside the (convex) boundary respectively.

Theorem 4.2. *Let Δ_{N+1} be the Delaunay triangulation obtained by inserting a point p into a Delaunay triangulation Δ_N of a point set P. Then all new triangles of Δ_{N+1} will have p as a common node.*

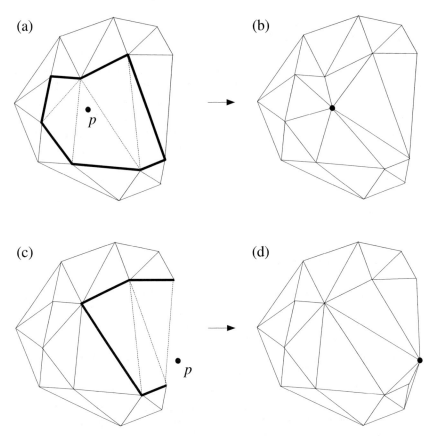

Fig. 4.3. Influence polygons Q^p in (a) and (c) when p is inserted interior and exterior to an existing triangulation. Delaunay triangulations after p has been inserted are shown in (b) and (d).

Proof. By Lemma 4.1, only the triangles whose circumcircle contain p will be modified by the insertion of p. Suppose that the theorem is false, that is, there is a Delaunay triangle $t_{i,j,k}$ of Δ_{N+1} such that its interior is strictly inside Q^p and p is not a node of $t_{i,j,k}$. The nodes p_i, p_j and p_k of $t_{i,j,k}$ are then Voronoi neighbors relative to the point set $\{P \cup p\}$. Then, by Lemma 3.5, they are also Voronoi neighbors relative to the subset P. Since the three Voronoi neighbors p_i, p_j and p_k span the Delaunay triangle $t_{i,j,k}$ in Δ_{N+1}, they also span a Delaunay triangle in Δ_N which must be $t_{i,j,k}$. But this is a contradiction since the circumcircles of the triangles of Δ_N in R^p contain p in their interior, so $t_{i,j,k}$ cannot be a Delaunay triangle of Δ_{N+1}. $\qquad\square$

It follows from the theorem that the nodes of the triangles of Δ_N in the influence region R^p must belong to the influence polygon Q^p. Further, all vertices of Q^p are visible from p when the old triangles of R^p have been removed. Then, the new triangulation Δ_{N+1} is obtained by removing all triangles of R^p and connecting p to all points of Q^p, see Figure 4.3. If R^p contains a single triangle, Δ_{N+1} is constructed trivially by connecting p to the three nodes of that triangle. If p lies outside Δ_N, p must also be connected to all nodes on the boundary of Δ_N that can be reached without crossing any edges of Δ_N. This ensures that a convex boundary is preserved for Δ_{N+1}, see Figure 4.3(d).

Another interesting task is the reverse of the point insertion operation, that of removing a point from a Delaunay triangulation. It is obvious that the theoretical results above apply directly to this problem. An influence region around the point to be removed must be found and retriangulated. Note that the influence polygon is not necessarily convex. The problem is studied in [41] and [62]. See also Exercises 6 and 7.

4.6 Point Insertion and Edge-Swapping

When using method (a) in Section 4.4 for incremental Delaunay triangulation, the search for triangles to be added to R^p can be done recursively around the insertion point p, without examining all triangles of Δ_N (unless all triangles of Δ_N are in R^p). Similarly, when using method (b), the swapping procedure can run recursively starting from the three initial edges incident with p without examining all edges of Δ_N, see Figure 4.4. The latter approach will now be examined in detail and we define an optimal algorithm for the swapping process to obtain the Delaunay triangulation Δ_{N+1}. Later we will explain how this algorithm is optimal. First recall the necessary steps to obtain Δ_{N+1}.

Step 1 Locate the triangle t in Δ_N that contains the insertion point p.

Step 2 Split t into three triangles by making three new edges between p and the nodes of t, and thus obtain a new triangulation Δ'_{N+1}.

Step 3 Apply a swapping procedure based on the circumcircle test (Section 3.7) to swap edges in Δ'_{N+1} until all edges are locally optimal and the final triangulation Δ_{N+1} is Delaunay.

Figure 4.5(a) shows the situation after point p has been inserted into the existing Delaunay triangulation Δ_N in Step 2, but before the proceeding step of swapping edges. The three new edges created in Step 2 cannot be swapped since they are diagonals of non-convex quadrilaterals, and by Definition 3.4 these edges are also locally optimal. The only edges that are *candidates* for swapping when starting the swapping procedure in Step 3, i.e., edges that may not be locally optimal and that may not pass the circumcircle test, are e_1, e_2 and e_3 in Figure 4.5(a). All other edges in Δ'_{N+1} are diagonals of

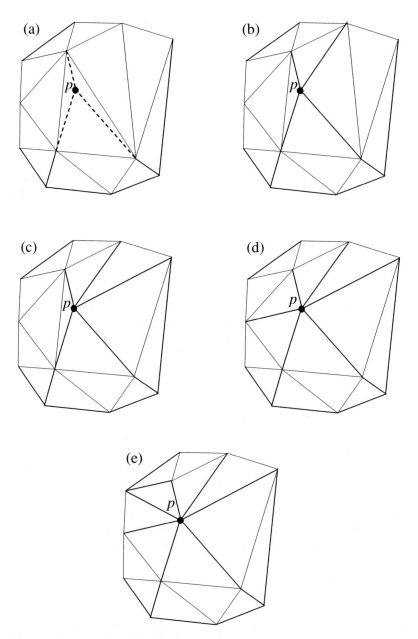

Fig. 4.4. Swapping procedure when inserting a point p into a Delaunay triangulation. From (b) to the final triangulation in (e), each picture shows the triangulation after one new edge has been swapped.

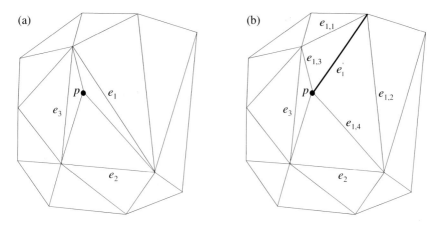

Fig. 4.5. Starting the recursive swapping procedure.

the same quadrilaterals as they were before inserting p into Δ_N, and since Δ_N is a Delaunay triangulation, those edges are locally optimal.

Assume that one of the edges e_1, e_2 or e_3, say e_1, is swapped to e_1' as a result of the circumcircle test, see Figure 4.5(b). Then the two edges $e_{1,1}$ and $e_{1,2}$ on the opposite side of e_1 from p become candidates for swapping since they become diagonals of new quadrilaterals and might not be locally optimal. In general, an edge-swap generates four new candidate edges for swapping since four new edges become diagonals of new quadrilaterals caused by the swap. Thus, in general, the two edges $e_{1,3}$ and $e_{1,4}$ incident with p should also be checked. However, when proceeding with swapping in Step 2 above, we will show in Lemma 4.2 below that when swapping an edge e_i, only the two edges $e_{i,1}$ and $e_{i,2}$ on the opposite side of e_i from p become candidates for swapping.

After e_1 has been swapped to e_1' in the example above, the edge $e_{1,1}$ passes the circumcircle test and is therefore locally optimal. But $e_{1,2}$ must be swapped and again generates two new candidates for swapping on the opposite side of $e_{1,2}$ from p. The situation when checking and swapping $e_{1,2}$ is exactly the same as for e_1: $e_{1,2}$ is a diagonal in a quadrilateral where two edges of the quadrilateral are incident with the insertion point p, and when $e_{1,2}$ is swapped it becomes incident with p. This argument can be repeated for the next candidate for swapping, and so forth, such that the swapping procedure fits into a binary tree structure in a recursive scheme where, by our hypothesis, each swapped edge generates exactly two new candidates for swapping. Figure 4.4(a)–(e) show all edge-swaps in the example above.

Let e_i be one of the edges e_1, e_2 or e_3 in Figure 4.5(a). Algorithm 4.3 takes e_i as input and swaps edges recursively as described above.

We now prove the hypothesis stated above that each edge-swap generates only two candidates for swapping, contrary to four in the general case.

Algorithm 4.3 recSwapDelaunay(Edge e_i)

---

1. **if** (circumcircleTest(e_i) == OK)   // Algorithm 3.1 in Section 3.7
2.   **return**
3. swapEdge(e_i)   // the swapped edge e'_i is incident with p
4. recSwapDelaunay($e_{i,1}$)   // call this procedure recursively
5. recSwapDelaunay($e_{i,2}$)   // call this procedure recursively

---

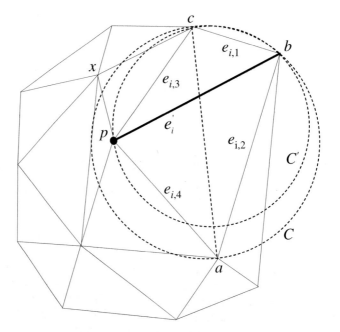

Fig. 4.6. Illustration for Lemma 4.2.

Lemma 4.2. *Each edge-swap in Algorithm 4.3 generates at most two candidate edges for swapping.*

Proof. Figure 4.6 shows the situation when an edge e_i (the dotted edge) has been swapped to e'_i in the example used previously. We must now prove that $e_{i,3}$ and $e_{i,4}$ are locally optimal and need not be considered for swapping. We prove this for $e_{i,3}$ by considering the quadrilateral (p, b, c, x) where $e_{i,3}$ is a diagonal. The proof for $e_{i,4}$ follows from symmetry.

The circle C passing through the nodes a, b and c has no other nodes than p in its interior since $t_{a,b,c}$ is a Delaunay triangle of Δ_N. Let C' be the circle passing through p, b and c, and let D' be the disc bounded by C'. Further let D_{cpb} denote the portion of D' which is bounded by the closed curve traced by the circle arc $(c - p - b)$ and the chord (b, c).

D_{cpb} is inside C since $(c - p - b)$ passes through p which is interior to C. But then D_{cpb} cannot contain any nodes in its interior, and in particular C' cannot contain x in its interior. This proves that $e_{i,3}$ is locally optimal.   □

Algorithm 4.3 is applied by starting with each of the edges, e_i, $i = 1, 2, 3$, and then swapping edges in each of the three angular sectors with node p as apex and spanned by e_i, $i = 1, 2, 3$. Each edge that is checked by the circumcircle test is on the opposite side of p in a triangle that has p as one of its nodes. An edge that is swapped will always become incident with p, and it will remain incident with p through the whole swapping process and will never be checked again. The following result now follows immediately.

Lemma 4.3. *No edge is checked more than once in Algorithm 4.3.*

Since all swapped edges are incident with p, it also follows that if an edge e_i passes the circumcircle test, the algorithm will never consider again any edge in the angular sector with vertex p as apex and spanned by e_i. It also follows from the lemma that the swapping procedure converges after a finite number of edge-swaps. The result is a triangulation Δ_{N+1} where all edges are locally optimal, and by Theorem 3.3 in Section 3.9, Δ_{N+1} must also be a Delaunay triangulation.

Since an edge that is checked is never considered again in Algorithm 4.3, it is also clear that an edge is not only locally optimal, but also Delaunay if it passes the circumcircle test or after it has been swapped. We formalize this with the following corollary.

Corollary 4.1. *The three initial edges created when inserting p are Delaunay edges. Similarly, any edge that is swapped (to p) is also a Delauney edge, as is any edge that passes the circumcircle test.*

This can also be verified using Theorem 3.1 in Section 3.4, by constructing a circle passing through the endpoints of the actual edge such that the circle contains no triangle nodes in its interior. Figure 4.7(a) shows the situation for the initial edges. The circumcircle C of the triangle t where p is located contains no other points than p in its interior since t is a Delaunay triangle in Δ_N. It is now possible to construct a circle C' passing through p and any of the nodes of t such that C' is tangent to C. Since C' is contained in C, it does not contain any points in its interior and the edge is Delaunay, by Theorem 3.1.

The same construction can be applied for an edge e_i that is swapped to p, see Figure 4.7(b). The circumcircle C of the old triangle t, on the opposite side of e_i from p, contains no other points than p in its interior since t is a Delaunay triangle of Δ_N. As in (a), we construct a circle C' passing through the endpoints of the swapped edge e_i' and tangent to C. Since C' is contained in C it does not contain any points in its interior, which proves that e_i' is Delaunay.

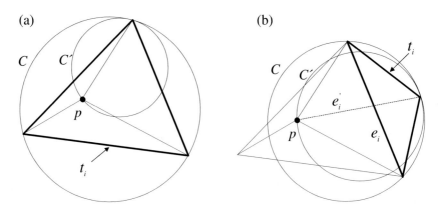

Fig. 4.7. (a): The initial edges are Delaunay. (b): Edges that are swapped to p are Delaunay.

Finally, we can also verify that any edge that passes the circumcircle test is Delaunay. This follows from the fact that the circle C in Figure 4.7(b) then would not contain p in its interior, and since t is Delaunay with respect to Δ_N, C contains no other nodes in its interior.

Lemma 4.3 and Corollary 4.1 justify in some sense our assertion in the introduction to this section that the swapping algorithm is optimal. Contrary to Lawson's local optimization procedure (LOP), where the same edge may be examined and swapped several times, Algorithm 4.3 examines every edge inside the influence region exactly once, and an edge that is swapped is never examined or swapped again. Since every edge is swapped to the insertion point p, we also get the following result.

Corollary 4.2. *The number of edge-swaps by* `recSwapDelaunay`, *when inserting a point p into a Delaunay triangulation, is* $\deg(p) - 3$, *where* $\deg(p)$ *is the degree of p in the new triangulation.*

It may happen that p falls on some existing edge e_i. If e_i is an interior edge, this makes no difference for Algorithm 4.3 as p is connected to the nodes of one of the two triangles sharing e_i. In such a case, there is a degenerate triangle along e_i when the swapping procedure starts, but the longer edge of that triangle will be swapped away by the algorithm. If e_i is a boundary edge, e_i must be split at p into two edges, and p is connected to the opposite node of e_i. The latter case need not be considered if the incremental algorithm starts with large enclosing triangles forming a triangulation with all points of the point set in its interior.

In Chapter 9, we put the different elements of incremental Delaunay triangulation together in a set of generic algorithms. In Particular, Section 9.7 gives a detailed recipe of how to program Step 1, Step 2 and Step 3 above for

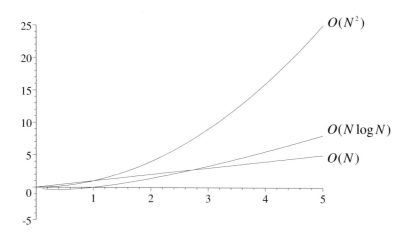

Fig. 4.8. The curves $y = x$, $y = x \log x$ and $y = x^2$ from bottom to top for illustrating the difference between run-time order of $O(N)$, $O(N \log N)$ and $O(N^2)$.

point insertion and edge-swapping generically using G-maps. Some examples of C++ code will also be given.

The theory of this section and the previous section was based on point insertion inside an existing Delaunay triangulation. Therefore, to construct the triangulation of a set of points P, we must start with an initial triangulation covering P such that each point can be inserted in the current triangulation's interior. But in practical applications we may also want to add points outside an existing triangulation. We leave to the reader to verify that the theory also holds when inserting points outside the convex hull of an existing Delaunay triangulation (Exercise 4).

4.7 Running Time of Incremental Algorithms

The following example by Lee & Schachter [52] shows that an incremental Delaunay triangulation algorithm is of order $O(N^2)$ in the worst case. (Consult Figure 4.8 to compare different complexities of algorithms.)

Consider $N = 10$ points on the parabola $y = \frac{1}{2}x^2$ as shown in Figure 4.9(a). Assume that the triangulation algorithm adds the points in the order they are numbered in the figure. For each point insertion, the triangulation is updated to be Delaunay. When inserting a new point p_i, the Delaunay criterion is violated for all existing triangles since the circumcircles of the triangles all contain p_i. Thus all existing triangles must be deleted. Insertion of p_i results in a "fan" of $(i - 1)$ new Delaunay edges radiating from p_i to all the points

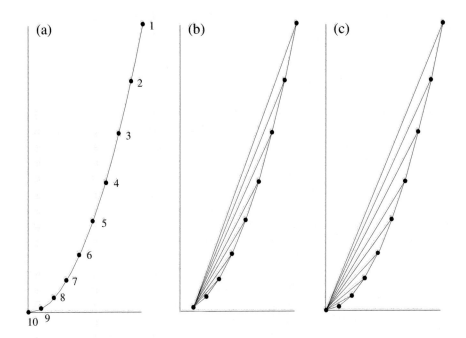

Fig. 4.9. Illustration of a worst case example for incremental Delaunay triangulation algoritms.

that have been inserted previously. Figure 4.9(b) shows the Delaunay triangulation after point p_9 has been inserted. When inserting p_{10}, the influence region consists of all triangles of the triangulation in (b) and consequently all existing triangles must be deleted. The new triangulation is shown in (c). In the worst case as shown in this example, the total work involved in updating the triangulation, counting new triangle edges, is $\sum_{i=4}^{N}(i-1) = \frac{1}{2}N^2 - \frac{1}{2}N - 3$, and this is of order $O(N^2)$.

Configurations of points like that of Figure 4.9 are very rare. In that example, after a point p_i has been inserted, the degree of that point is $i - 1$ in the updated Delaunay triangulation. But, from (1.9) in Section 1.3 we see that the average degree of a vertex is less than six. Combining this with Corollary 4.2 above and assuming that the points are added in random order, it can be shown that the *expected* number of edge-swaps to insert a point is at most three when using Algorithm 4.3. Moreover, together with an efficient search structure for localizing the triangle containing the insertion point (implemented as a directed acyclic graph), it can be shown that the expected time for constructing a Delaunay triangulation of N points is $O(N \log N)$.

Details of this probabilistic analysis can be found in [19] and [26]. This is also known as *randomized incremental* Delaunay triangulation.

4.8 Divide-and-Conquer

Divide-and-conquer is a general technique used to solve many problems in computational geometry. We pay some attention to this technique since the divide-and-conquer algorithm is the only one, besides Fortune's sweepline algorithm [31], that has been proved to have a theoretical running time of order $O(N \log N)$. (Revisit Figure 4.8 to compare run-time order of algorithms.) Divide-and-conquer is static like the Radial Sweep algorithm and the step-by-step approach in Section 4.3. It treats all points of a point set P simultaneously, and a valid Delaunay triangulation of P is achieved in the last step of the algorithm. Divide-and-conquer is more involved and more difficult to implement than most other algorithms.

The basic idea is to recursively subdivide the point set into two data sets of approximately equal size until each data set contains only a small number of points. Let $P = P_0^0$ be the original point set, and let the subsets P_0^1 and P_1^1 be the result from the first split of P at level $k = 1$. We may assume here that P_0^1 lies to the left of P_1^1 in the plane, that is, the x-values of points in P_0^1 are less then those of P_1^1. The following is an illustration of the recursive splitting of P over the first three levels, where a subset P_i^k at level k is split into two subsets P_{2i}^{k+1} and P_{2i+1}^{k+1} at the next level $k + 1$.

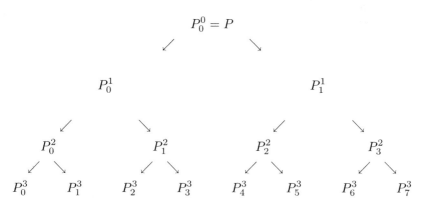

Delaunay triangulations, each containing only a few triangles, are constructed from the point sets at the lowest level. Pairs of triangulations are then merged such that each merged triangulation is Delaunay. That is, the Delaunay criteria are fulfilled and the boundary of each merged triangulation is convex. This process goes on until the two triangulations Δ_0^1 and Δ_1^1 of

P_0^1 and P_1^1 at the first subdivision level are merged to obtain the Delaunay triangulation $\Delta(P)$ of the whole point set:

$$
\underbrace{\Delta_0^3 \qquad \Delta_1^3} \qquad \underbrace{\Delta_2^3 \qquad \Delta_3^3} \qquad \underbrace{\Delta_4^3 \qquad \Delta_5^3} \qquad \underbrace{\Delta_6^3 \qquad \Delta_7^3}
$$

$$
\underbrace{\Delta_0^2 \qquad\qquad\qquad \Delta_1^2} \qquad \underbrace{\Delta_2^2 \qquad\qquad\qquad \Delta_3^2}
$$

$$
\underbrace{\Delta_0^1 \qquad\qquad\qquad\qquad\qquad\qquad \Delta_1^1}
$$

$$
\Delta_0^0 = \Delta(P)
$$

Apart from the merge process, the steps of the algorithm are trivial. A detailed description and analysis of different merging algorithms for Delaunay networks can be found in [61]. We describe briefly the merge approach used in [52] and [34] omitting details on data structures used by the algorithms. The N points of P are first sorted in lexicographically ascending order such that $p_i = (x_i, y_i) < (x_j, y_j) = p_j$ if and only if $x_i < x_j$, or $x_i = x_j$ and $y_i < y_j$. Next, P is subdivided into two subsets P_L and P_R such that $P_L = \{p_1, \dots, p_{\lfloor N/2 \rfloor}\}$ and $P_R = \{p_{\lfloor N/2 \rfloor + 1}, \dots, p_N\}$. The Delaunay triangulations $\Delta(P_L)$ and $\Delta(P_R)$ are now constructed recursively and merged to obtain a Delaunay triangulation $\Delta(P_L \cup P_R)$. The merging of $\Delta(P_L)$ and $\Delta(P_R)$ consists of two steps, see Figure 4.10.

1. The convex hull of $\Delta(P_L \cup P_R)$ is determined. Since we already have computed the convex hull of $\Delta(P_L)$ and $\Delta(P_R)$, this can be done by finding the lower and upper common tangents e_b and e_u of the two triangulations as indicated in Figure 4.10(a). Each tangent connects a node in $\Delta(P_L)$ with a node in $\Delta(P_R)$. Since these nodes are on the convex hull of $P_L \cup P_R$, the tangents define Delaunay edges in $\Delta(P_L \cup P_R)$. Note that the nodes of e_u and e_b are not necessarily those with maximum and minimum y-value in the two triangulations.
2. The merging starts at the lower common tangent e_b, the base line, which has endpoints p_L and p_R. Then a point from $\Delta(P_L)$ or $\Delta(P_R)$ is looked for to form a Delaunay triangle with e_b. This is done exactly as in the step-by-step approach in Section 4.3. A circle passing through p_L and p_R "grows" upward until a point p is reached. The points p_L, p_R and p now form a Delaunay triangle of $\Delta(P_L \cup P_R)$ since there are no points from $P_L \cup P_R$ interior to its circumcircle (Theorem 4.1). Edges of $\Delta(P_L)$ and $\Delta(P_R)$ that are intersected by the new edge are removed as indicated by the dotted

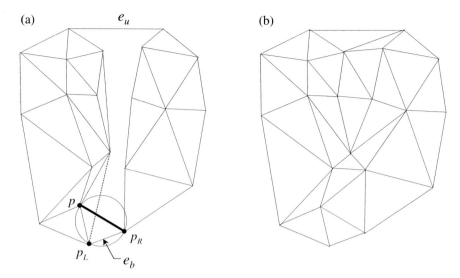

Fig. 4.10. Merging two Delaunay triangulations in the divide-and-conquer algorithm.

edge in $\Delta(P_L)$. The inserted edge is set to the new base line and the same procedure is applied repeatedly until the upper common tangent e_u is reached. The final triangulation $\Delta(P_L \cup P_R)$ shown in Figure 4.10(b) must be Delaunay since no points of $\{P_L \cup P_R\}$ are interior to any circumcircle of a triangle in the triangulation.

Let $t(N)$ denote the time for constructing the Delaunay triangulation of a set of N points. Then we have the following recurrence relation;

$$t(N) = 2t(N/2) + M(N/2, N/2) \tag{4.1}$$

where $M(N/2, N/2)$ is the time needed to merge two triangulations with $N/2$ nodes in each. The merge process, if properly implemented, can be shown to be linear in time. Thus, the term $M(N/2, N/2)$ is of order $O(N)$. (See Exercise 8).

Theorem 4.3. *The divide-and-conquer algorithm is of order $O(N \log N)$.*

Proof. Assume that, for some integer k, the point set P was divided into 2^k subsets. Expanding the recurrence relation (4.1), we get

$$t(N) = 2t(N/2) + M(N/2, N/2) = 2(2t(N/4) + M(N/4, N/4)) + M(N/2, N/2)$$
$$= 4t(N/4) + 2M(N/2, N/2)$$
$$= 8t(N/8) + 3M(N/2, N/2)$$
$$= \cdots$$
$$= 2^k t(N/2^k) + kM(N/2, N/2).$$

The first term in the last expression represents the time needed to create 2^k triangulations at the lowest level, and the second term represents the total time of all the merge processes. Without loss of generality, assume that the number of points in P is a power of 2, say $N = 2^m$. Since there are only a few points in each subset at the lowest level, we have $m \approx k$ when N is large. If Delaunay triangulation is of order $O(N^2)$, cf. Section 4.7, then the first term is of order $2^k (N/2^k)^2 = 2^{2m-k} \approx 2^m = O(N)$. With the assumption above that the merge process can be done in linear time, the second term is of order $O(kN) \approx O(mN)$. Since $m = \log_2 N$, we get $O(N \log N)$ for the merge processes, which we conclude is the total complexity of the divide-and-conquer algorithm. (The initial lexicographic sorting can also be implemented in $O(N \log N)$ time.)   □

It has been shown that the minimal complexity for computing the convex hull of a set of N points in the plane is $O(N \log N)$ [68]. Since Delaunay triangulation of a set of points also computes the convex hull of the points, the divide-and-conquer algorithm is in fact optimal. But, as pointed out in Section 4.7, the average performance in practical usage relies heavily on the underlying data structure and aspects of implementation, so the divide-and-conquer algorithm need not necessarily perform better than other algorithms.

Divide-and-conquer techniques are also used for constructing the Voronoi diagram of a set of points, see for example [76] and [68], and as shown in Section 3.4, the Delaunay triangulation is easily derived from the Voronoi diagram.

4.9 Exercises

1. Let p_i, p_j and p_k be strong Voronoi neighbors in a point set P. (p_i, p_j and p_k are not the only points in P). Is $t_{i,j,k}$ necessarily a Delaunay triangle of the Delaunay triangulation of P? Make a figure with the Voronoi diagram and the Delaunay triangulation and explain.
2. Let p_j be the closest point to p_i in a point set P, and let e_{ij} be an edge with p_i and p_j as endpoints. Why is e_{ij} a Delaunay edge with respect to P?
3. Prove Theorem 4.1.

4. Verify that the theory of Section 4.5 and 4.6 also holds for point insertion outside the convex hull of an existing Delaunay triangulation (with minor modifications).

5. Design and implement an algorithm for Delaunay triangulation by the step-by-step approach in Section 4.3.

6. Establish the theory for removing a node from a Delaunay triangulation. Hint: Regard the problem as the reverse operation of inserting a point in Section 4.5, but do not expect to find an algorithm without using an LOP-procedure.

7. Design two algorithms for removing a point p from a Delaunay triangulation Δ_N to obtain a triangulation Δ_{N-1} that is also Delaunay.

   a) p is interior to Δ_N.

   b) p is on the (convex) boundary of Δ_N.

8. Let N be the total number of edges in two triangulations $\Delta(P_L)$ and $\Delta(P_R)$ that are merged in the divide-and-conquer algorithm in Section 4.8. Show that the merge process can be done in $O(N)$ time:

   a) Show that finding the lower and upper tangents e_b and e_u of $\Delta(P_L)$ and $\Delta(P_R)$ can be done in $O(N)$ time when the convex hulls of $\Delta(P_L)$ and $\Delta(P_R)$ are present as two closed polygons.

   b) Show that step 2 of the merge step is $O(N)$ (after the lower and upper tangents e_b and e_u have been found).

   Assume that the underlying data structure has sufficient adjacency information to carry out topological queries efficiently.

5

Data Dependent Triangulations

Although Delaunay triangulations have well-shaped triangles in the plane and satisfy an optimum principle by the MaxMin angle criterion, they are not necessarily optimal as domains for surfaces. In this chapter we are concerned with surfaces in 3D space defined over triangulations, and in particular surfaces represented by piecewise linear functions over triangulations in the plane. The main concern in this respect is to construct triangulations from 3D point sets where the triangulations are optimal according to local and global cost functions which are designed to reflect properties of the underlying physical model from which the points have been sampled.

5.1 Motivation

Suppose that a triangulation $\Delta(P)$ of a set of points P in the plane is given, and assume that each point $p_i = (x_i, y_i)$, $i = 1, \ldots, N$ in P has an associated real value z_i. We assume that the data $\{(x_i, y_i, z_i)\}$ are samples from some underlying function, or surface, F. Let f_Δ be an approximation to F that belongs to the function space $S_1^0(\Delta)$ defined in Chapter 1. That is, f_Δ is the unique surface triangulation which is a linear polynomial over each triangle t_i in Δ, and which interpolates the data $\{(x_i, y_i, z_i)\}$,

$$f_\Delta|_{t_i} \in \Pi_1, \tag{5.1}$$
$$f_\Delta(x_i, y_i) = z_i, \quad i = 1, \ldots, N,$$

where Π_1 is the space of linear polynomials.

It is clear that f_Δ depends on the particular choice of the triangulation Δ. In the following, we illustrate this by considering piecewise linear approximations to a smooth test function $F_1(x, y) = (\tanh(9y - 9x) + 1)/9$ taken from Franke [32] (Figure 5.1). The function represents a sharp rise running

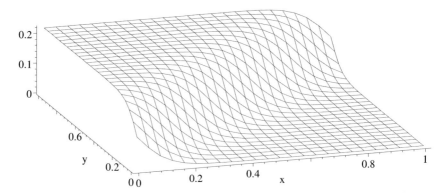

Fig. 5.1. Test function $F_1(x, y) = (\tanh(9y - 9x) + 1)/9$.

diagonally across the unit square. Note that the first and second directional derivatives of F_1 in the diagonal direction $u = (1, 1)$ are zero everywhere.

Figure 5.2(a) shows a Delaunay triangulation of a point set chosen on a regular grid in the unit square[1], and (b) shows the level curves of the approximation f_Δ to F_1. Assume that we construct another triangulation Δ' of the same point set and define the function $f_{\Delta'}$ over Δ'; see Figure 5.3(a) and (b). The level curves of $f_{\Delta'}$ are clearly more exact than those of f_Δ and they indicate that $f_{\Delta'}$ is more smooth, in some sense, than f_Δ. In many applications we would prefer $f_{\Delta'}$ to f_Δ and thus the triangulation Δ' to Δ, even though Δ' does not possess the "equiangular" property of Delaunay triangulations. Note that many triangles in Δ' are long in directions where the magnitude of the second directional derivative of the underlying function F_1 is small, and they are narrow in directions where the magnitude of the second directional derivative is large. In many cases, this property is desirable as it may yield better approximations of smooth functions with linear polynomials [71, 25].

The observations above suggest that we should seek alternatives to the Delaunay criteria for deciding if a triangulation is optimal. In particular we should look for criteria that use the data values $\{z_i\}$.

5.2 Optimal Triangulations Revisited

Optimal triangulations were studied in Chapter 3 in the context of triangulations that were optimal according to the Delaunay criteria. Delaunay triangulations had the remarkable property that if all interior edges in a triangulation Δ were *locally* optimal according to the MaxMin angle criterion,

---

[1] The Delaunau triangulation in Figure 5.2(a) is not unique as all the diagonals in the grid cells represent neutral cases. The directions of the diagonals are chosen randomly.

(a)   (b)

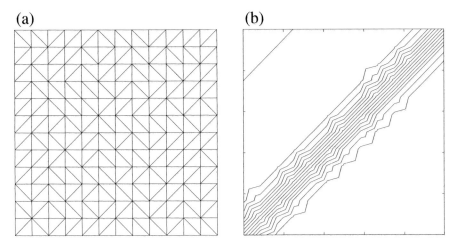

Fig. 5.2. Delaunay triangulation of grid data and level curves.

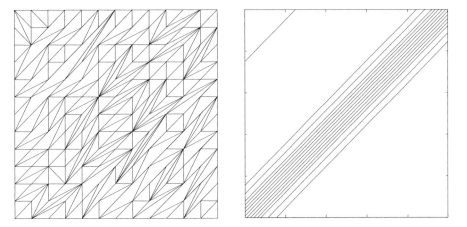

Fig. 5.3. Triangulation optimized with the LOP algorithm using ABN criterion and l_1 norm.

or other equivalent Delaunay criteria defined in the plane, then Δ was also *globally* optimal as shown in Section 3.9. Recall the basis from the theoretical study that led to this optimality consideration:

1. A local cost α_i was associated with each triangle t_i in a triangulation Δ^k, where α_i was the angular measure of the smallest interior angle in t_i.
2. The interior angles α_i were ordered non-decreasingly in an indicator vector $I(\Delta^k) = (\alpha_1, \alpha_2, \ldots, \alpha_{|T|})$. Two triangulations Δ^k and Δ^l were compared according to the lexicographical ordering of their indicator vectors. We said that Δ^k was "better" than Δ^l if $I(\Delta^k)$ was lexicographically larger

than $I(\Delta^l)$. Thus, the lexicographical measure of an indicator vector defined a global cost function.

3. The optimal triangulation among all possible triangulations of a point set was the triangulation with the indicator vector which was lexicographically largest.

The Local Optimization Procedure (LOP) was introduced in Section 3.8 and applied to an arbitrary triangulation to obtain a locally optimal triangulation that was also globally optimal. This was done by swapping interior edges which were diagonals of convex quadrilaterals in the triangulation. We defined the resulting triangulation as a Delaunay triangulation and we showed in Section 3.9 that it was also a Delaunay triangulation as defined by the straight line dual of the Voronoi diagram.

This concept will be generalized in several ways in the following. We will consider other local cost functions than that which was based on the MaxMin angle criterion for Delaunay triangulations. In particular the local cost functions will be defined using the data values $\{z_i\}$ giving rise to so-called *data dependent triangulations*. Further, we will use global measures of the cost functions other than the lexicographical measure. With these extensions we lose an important property possessed by Delaunay triangulations: a locally optimal triangulation will in general not be globally optimal. In fact, the MaxMin angle criterion is the only known criterion with this property. So, applying the LOP with criteria other than the MaxMin angle criterion delivers a triangulation that may be stuck in a local optimum which is not a global optimum in general. Therefore, other optimization procedures in addition to the LOP will be considered in an attempt to obtain a global cost function that is closer to a global optimum than obtained by the LOP.

5.3 The General Concept

Given an arbitrary triangulation $\Delta(P)$ of the set of points P with associated data values $\{z_i\}$, and a surface triangulation f_Δ over Δ as defined in (5.1). To each interior edge e_i of Δ we associate a local cost function $c(\Delta, e_i)$. In general the local cost associated with an interior edge e_i will be some measure of the sharpness of the surface triangulation along that edge, for example, the angle between the normals of the two triangles in 3D space sharing e_i. Several local cost functions are introduced and compared in the next sections. An indicator vector is constructed with the local costs as entries,

$$I(\Delta) = \left(c(\Delta, e_1), \ldots, c(\Delta, e_{|E_I|})\right). \tag{5.2}$$

The *quality* of the triangulation Δ is now determined by measuring the indicator vector in some appropriate norm. A natural approach is to use the discrete l_1 and l_2 norms which give rise to the following real valued *global cost functions* of a triangulation,

$$C_p(\Delta) = \sum_{i=1}^{|E_I|} |c(\Delta, e_i)|^p, \quad p = 1, 2. \tag{5.3}$$

Different triangulations of the same point set P can now be compared by comparing their associated global cost functions. Recall that two triangulations of the same point set have the same number of triangles and interior edges if they have the same external boundary; cf. Lemma 1.1. The *optimal triangulation* among all possible triangulations of P will be the triangulation with the smallest global cost function. Since the local cost functions measure sharpness of the triangulation along its edges, the optimal triangulation will be overall smooth in some sense. This is explained in more detail later.

One could also use a global cost function based on the l_∞ norm, $C_\infty(\Delta) = \max_{E_i} |c(\Delta, e_i)|$, or the global cost function can be based on lexicographical ordering as in Section 3.1. If we assume that the indicator vectors of two triangulations Δ and Δ' are ordered non-increasingly, then the query $I(\Delta') < I(\Delta)$ by the lexicographical ordering scheme agrees with $C_\infty(\Delta') < C_\infty(\Delta)$ if the first entry of the two indicator vectors differ. Thus, a global cost function based on the lexicographical ordering scheme can be regarded as a refined l_∞ norm.

The effect of swapping an edge of a convex quadrilateral in a triangulation Δ can now be read from the actual cost function used. In the data dependent swapping criteria introduced below, the local cost associated with an interior edge e_i of a triangulation Δ is calculated using only the geometric embedding information of the two triangles sharing e_i. Thus, the effect of swapping e_i to e_i' and transforming Δ to a new triangulation Δ', can be checked locally. In addition to the local cost associated with e_i, the cost functions associated with the four edges, e_i^k, $k = 1, 2, 3, 4$, sharing the triangles t_k and t_l with e_i must also be checked, see Figure 5.4. Using the discrete l_p norms with $p = 1$ or $p = 2$, the following inequality must be evaluated to decide if Δ is "better" than Δ',

$$|c(\Delta, e_i)|^p + \sum_{k=1}^{4} |c(\Delta, e_i^k)|^p < |c(\Delta', e_i')|^p + \sum_{k=1}^{4} |c(\Delta', e_i^k)|^p. \tag{5.4}$$

If this inequality holds, then the global cost function in (5.3) associated with Δ is strictly smaller than the cost function associated with Δ', and Δ is "more optimal" than Δ'. We call e_i a *locally optimal edge* in Δ if (5.4) holds or if the quadrilateral with e_i as a diagonal is not strictly convex so that e_i cannot be swapped. Further, if all interior edges in Δ are locally optimal, then Δ will be called a *locally optimal triangulation* . This implies that Δ is locally optimal if the real valued function $C_p(\Delta)$ cannot be reduced by swapping any edge

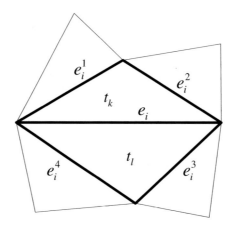

Fig. 5.4. Edges that must be checked when considering swapping the edge e_i.

in Δ. These definitions are similar to those of Section 3.8 for the Delaunay swapping criteria. The local optimization procedure in Section 3.8 can now be applied with more general swapping criteria.

Algorithm 5.1 LOP for data dependent triangulations

---

1. Make an arbitrary legal triangulation Δ of a point set P.
2. **if** Δ is locally optimal, that is, if (5.4) holds for all interior edges in Δ,
   stop.
3. Let e_i be an interior edge of Δ which is not locally optimal.
4. Swap e_i to e_i', transforming Δ to Δ'.
5. Let $\Delta := \Delta'$.
6. **goto** 2.

The only difference between Algorithm 5.1 and Algorithm 3.2 is Step 2 and Step 3 which allow for arbitrary swapping criteria as those introduced in the next section. A natural guess for an initial triangulation would be a Delaunay triangulation which can be generated efficiently with one of the algorithms outlined in Chapter 4. Each time an edge-swap occurs, the global cost represented by the real valued function (5.3) decreases, and since the number of possible triangulations of P is finite, the algorithm converges after a finite number of edge-swaps. On termination of the algorithm, all interior edges are locally optimal and (5.3) has reached a local minimum. Some notes on implementation of the LOP are given in Section 5.5.

Note again that, when using general swapping criteria, Algorithm 5.1 does not necessarily deliver a triangulation that is globally optimal as was the case with the Delaunay swapping criteria. There might still be edges with high local costs in the final triangulation since they are interior to non-convex quadrilaterals and thus not swappable. It is also important to note that the triangulation produced by Algorithm 5.1 may depend on the specific order in which edges are examined and swapped in Step 3 and 4 of the algorithm. This is also a major difference from using Delaunay criteria for edge-swapping as a Delaunay triangulation produced by the LOP is unique apart from neutral cases.

5.4 Data Dependent Swapping Criteria

In this section a class of edge-swapping criteria is introduced that can be used in the LOP algorithm and in other algorithms outlined later. The swapping criteria are called *data dependent* since the function values $\{z_i\}$ at the nodes are used when considering edge-swaps. This is in contrast to Delaunay swapping criteria, which are based on information about the position of the nodes in the plane only. Consequently, the triangulations produced using these criteria are called *data dependent triangulations*. More specifically, we design local cost functions $c(\Delta, e_i)$ associated with each edge e_i of a triangulation which are used in the global cost functions in (5.3).

The local cost associated with an edge e_i is calculated using the 3D geometric embedding information of the two triangles t_1 and t_2 sharing e_i, see Figure 5.5. For common notation and comparison of the local cost functions presented below, we use the two planes Q_1 and Q_2 associated with t_1 and t_2,

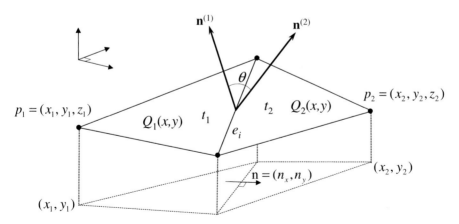

Fig. 5.5. Geometric embedding information used when considering edge-swap of an edge e_i in a convex quadrilateral defined by two triangles t_1 and t_2. $Q_1(x,y)$ and $Q_2(x,y)$ are the equations of the planes defined by t_1 and t_2. The dashed lines display the projection of the quadrilateral in the (x,y) -plane. $\mathbf{n} = (\mathbf{n}_x, n_y)$ is a unit vector orthogonal to the projection of e_i in the (x,y)-plane.

$$Q_i(x, y) = a_i x + b_i y + c_i, \quad i = 1, 2.$$

That is, Q_1 and Q_2 are the restrictions of the piecewise linear surface triangulation f_Δ to t_1 and t_2 respectively.

Angle between normals (ABN). Let $\mathbf{n}^{(1)}$ and $\mathbf{n}^{(2)}$ be normal vectors to the two planes Q_1 and Q_2 respectively, see Figure 5.5. The normal vectors are the gradients of the implicit forms $q_i(x, y, z) = -a_i x - b_i y - c + z = 0$, $i = 1, 2$, of Q_1 and Q_2,

$$\mathbf{n}^{(i)} = \nabla q_i(x, y, z) = \left(\frac{\partial q_i}{\partial x}, \frac{\partial q_i}{\partial y}, \frac{\partial q_i}{\partial z} \right) = (-a_i, -b_i, 1), \quad i = 1, 2.$$

The ABN cost function is defined as the acute angle θ between $\mathbf{n}^{(1)}$ and $\mathbf{n}^{(2)}$ as shown in Figure 5.5, and θ can be expressed by the following scalar product (see also Section 3.7),

$$\cos \theta = \frac{\mathbf{n}^{(1)} \cdot \mathbf{n}^{(2)}}{\left\| \mathbf{n}^{(1)} \right\|_2 \left\| \mathbf{n}^{(2)} \right\|_2}.$$

Expressed by the plane equations, we get the local cost function

$$c_{ABN}(\Delta, e_i) = \theta = \cos^{-1} \frac{\mathbf{n}^{(1)} \cdot \mathbf{n}^{(2)}}{\left\| \mathbf{n}^{(1)} \right\|_2 \left\| \mathbf{n}^{(2)} \right\|_2} = \cos^{-1} \frac{a_1 a_2 + b_1 b_2 + 1}{\sqrt{(a_1^2 + b_1^2 + 1)(a_2^2 + b_2^2 + 1)}}.$$

When considering a quadrilateral with an edge e_i as a diagonal, the cost function measures the angle between the two planes defined by the triangles sharing e_i, or in other words, the sharpness, or the local smoothness along the edge e_i in the triangulation Δ. If the quadrilateral is strictly convex, e_i can be swapped to e_i' transforming Δ to a new triangulation Δ'. Replacing $c(\Delta, e_i)$ with $c_{ABN}(\Delta, e_i)$ in (5.3), or evaluating (5.4), the effect of the swapping can be measured globally. Running Algorithm 5.1 with c_{ABN} as the local cost function produces a triangulation where all interior edges are locally optimal. Since c_{ABN} measures smoothness in some sense, the resulting triangulation at termination of the LOP has also reached a local optimal state of smoothness in some sense.

The triangulation in Figure 5.3 was generated with c_{ABN} as the local cost function. Algorithm 5.1 was used with the Delaunay triangulation in Figure 5.2 as the initial triangulation. The discrete l_1 norm (with $p = 1$ in (5.3)) was used to measure the global cost.

Jump in normal derivative (JND). An alternative measure of sharpness of the edge e_i is the jump in the normal derivatives of Q_1 and Q_2 over the edge e_i. Let $\mathbf{n} = (n_x, n_y)$ be a unit vector in the xy-plane orthogonal to the projection of e_i in the xy-plane. (See Figure 5.5). The derivative of Q_i in the direction of $\mathbf{n}$ is defined as the scalar product,

$$\partial Q_i / \partial \mathbf{n} = \nabla Q_i \cdot \mathbf{n} = (\partial Q_i / \partial x, \ \partial Q_i / \partial y) \cdot (n_x, n_y) = (a_i n_x + b_i n_y).$$

The cost function can be written as,

$$c_{JND}(\Delta, e_i) = |\partial Q_1 / \partial \mathbf{n} - \partial Q_2 / \partial \mathbf{n}| = |(a_1 - a_2)n_x + (b_1 - b_2)n_y|. \quad (5.5)$$

This measure can also be regarded as a second order divided differences operator. In Section 8.6 we use it to construct a smoothing term involved in a least squares surface approximation of scattered data over triangulations. An explicit expression for (5.5) is also derived there.

Deviations from linear polynomials (DLP). The general form of this cost function is

$$c_{DLP}(\Delta, e_i) = \left\| \begin{pmatrix} |Q_1(x_2, y_2) - z_2| \\ |Q_2(x_1, y_1) - z_1| \end{pmatrix} \right\|.$$

The cost function measures the vertical distance between the plane Q_1 and the vertex p_2, and vice versa for Q_2 and p_1, in some norm $\|\cdot\|$ defined on R^2. A natural approach is to use an l_p norm,

$$c_{DLP}(\Delta, e_i) = (|Q_1(x_2, y_2) - z_2|^p + |Q_2(x_1, y_1) - z_1|^p)^{1/p}$$

where $p = 1$ or $p = 2$.

Distance from planes (DFP). The vertical distances in the DLP cost function above are now replaced by the normal distances from the points p_2 and p_1 to the planes Q_1 and Q_2 respectively. The general form is,

$$c_{DFP}(\Delta, e_i) = \left\| \begin{pmatrix} \text{dist}(Q_1, p_2) \\ \text{dist}(Q_2, p_1) \end{pmatrix} \right\|.$$

The distance between a plane $Q(x, y) = ax + by + c$ and a point $p_l = (x_l, y_l, z_l)$ is given by $\text{dist}(Q, p_l) = |Q(x_l, y_l) - z_l| / (a^2 + b^2 + 1)^{1/2}$. Again, a natural choice is to use an l_p norm,

$$c_{DFP}(\Delta, e_i) = (\text{dist}(Q_1, p_2)^p + \text{dist}(Q_2, p_1)^p)^{1/p}$$

with $p = 1$ or $p = 2$.

Smoothness of contours (SCO). A surface triangulation is frequently presented as a contour map consisting of level curves that are intersection curves between horizontal planes and the surface triangulation. If the surface triangulation is represented as piecewise linear surface patches, then the level

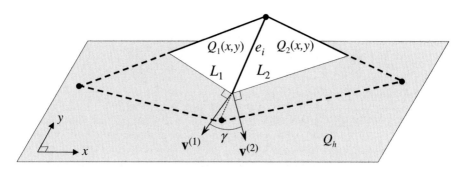

Fig. 5.6. The SCO data dependent swapping criterion.

curves are also piecewise linear. This cost function is designed to minimize spikes in the level curves and thus to obtain pleasant looking contour maps.

Let the two planes Q_1 and Q_2 be intersected by a horizontal plane Q_h as shown in Figure 5.6. The (horizontal) intersection lines are indicated as L_1 and L_2. Let $\mathbf{v}^{(1)}$ and $\mathbf{v}^{(2)}$ be horizontal vectors orthogonal to L_1 and L_2 as shown in the figure. It can be shown that $\mathbf{v}^{(1)}$ and $\mathbf{v}^{(2)}$ have the same directions as $\mathbf{n}^{(1)}$ and $\mathbf{n}^{(2)}$ respectively, under the ABN cost function above when $\mathbf{n}^{(1)}$ and $\mathbf{n}^{(2)}$ are projected onto a horizontal plane (Exercise 1). Hence, $\mathbf{v}^{(1)} = (-a_1, -b_1, 0)$ and $\mathbf{v}_2 = (-a_2, -b_2, 0)$. Let the cost function measure the angle γ between $\mathbf{v}^{(1)}$ and $\mathbf{v}^{(2)}$,

$$c_{SCO}(\Delta, e_i) = \gamma = \cos^{-1} \frac{\mathbf{v}^{(1)} \cdot \mathbf{v}^{(2)}}{\left\| \mathbf{v}^{(1)} \right\|_2 \left\| \mathbf{v}^{(2)} \right\|_2} = \cos^{-1} \frac{a_1 a_2 + b_1 b_2}{\sqrt{(a_1^2 + b_1^2)(a_2^2 + b_2^2)}}.$$

This cost function gives the same measure as $c_{ABN}(\Delta, e_i)$ if e_i is vertical. But rotating the two triangles which share e_i until e_i is horizontal, we see that the angle γ increases while the spatial angle θ of the $c_{ABN}(\Delta, e_i)$ cost function is not affected.

Roughly speaking, the rationale for this cost function is to avoid sharp edges in flat (horizontal) areas, thus producing visually pleasant contour maps. The triangulation in Figure 5.12 was generated with c_{SCO} as the local cost function. The level curves are clearly more smooth and pleasant looking than those from the Delaunay triangulation in Figure 5.11.

All the local cost functions above represent sharpness of an edge in a triangulation in some sense. They all evaluate to zero if $Q_1 = Q_2$, that is, if the two triangles t_1 and t_2 shared by e_i lie in the same plane. This corresponds to a neutral case for which swapping of e_i does not change the global cost. The JND and DFP criteria are only defined for surface triangulations that are functions of two variables, while the other criteria apply for general surface triangulations in 3D space.

The first four cost functions were tested by Dyn et al. [25]. They found that the ABN and JND criteria were in general the best choice for approximating a selection of smooth test functions with piecewise linear surface triangulations. Contour maps derived from triangulations optimized with the SCO cost function are in general smoother than contour maps derived from triangulations optimized with other criteria. It should be mentioned, however, that in many cases there are only minor visual differences between triangulations generated using different data dependent criteria. Also, a criterion that works well on one data set may perform worse and behave differently on another data set. The ABN criterion was used by Choi et al. [16] for improving an initial triangulation that was constructed from scattered data in 3D space. They used a slightly different approach from the concept outlined above as they did not include the diagonal edge e_i when considering edge-swaps by (5.4).

5.5 On Implementation of the LOP

The local optimization procedure (LOP) is easy to implement and it is extremely fast if sufficient topological information is available in the data structure to perform edge-swaps efficiently. The only operation of Algorithm 5.1 that modifies the topology of the triangulation is the swapping in Step 4 which must be implemented on the actual data structure. The other operations are simple topological queries and geometric calculations for computing the local cost of interior edges using the swapping criteria of Section 5.4.

Starting with an initial triangulation, for example a Delaunay triangulation, all the interior edges in the triangulation can be maintained in a linear array which is passed through several times by Algorithm 5.1 until no edge-swaps occur. In each iteration of the LOP, from Step 2 to Step 6, all edges are examined in the same sequence as the ordering in the array. Each entry in the list must have sufficient information for finding geometric embedding information needed for calculating the local cost and for finding topological neighbors of the edge in the entry.

An alternative to maintaining all interior edges in the list is to store only those edges that are *not* locally optimal, that is, those edges that are candidates for swapping. After each swap of an edge e_i, this edge will be removed from the list since it becomes locally optimal. In addition, the edges of the triangles t_k and t_l sharing e_i must be checked, see Figure 5.4. Edges that change from being locally optimal to become not locally optimal, or vice versa, will now be added or removed from the list.

Note that the indicator vector (5.2) need not be maintained and stored explicitly. Steps 3 and 4 of Algorithm 5.1 only require a test which decides if an edge should be swapped using the inequality (5.4); calculating the global cost (5.3) is not required by the LOP.

In most cases, the LOP algorithm converges after only a few passes through the list of edges terminating with the global cost function in (5.3) at a local minimum. But on termination of the algorithm there might be many edges with relatively high costs that are not swappable in the final triangulation since they are interior to non-convex quadrilaterals. These edges might cause undesirable visual effects and contribute to large approximation errors.

5.6 Modified Local Optimization Procedures (MLOP)

Since the triangulation produced by Algorithm 5.1 may depend on the specific order in which edges are swapped, it is natural to look for strategies to control the order in which this is done. One such strategy, proposed in [24], is to swap, in each iteration of Algorithm 5.1, the edge which gives the maximum reduction of the global cost function. Another strategy is to swap the edge which will leave the maximum possible number of edges swappable for the next iteration of the LOP.

Numerical experiments do not indicate significant improvements using these strategies compared to using the standard LOP algorithm. The first strategy, even though it may converge in a lower number of edge-swaps than the LOP, will in many cases deliver a triangulation with a global cost function that is greater than that obtained by the LOP. This is probably caused by the fact that many edges with high local cost are "frozen" at an optimal state too early in the process thus preventing neighboring edges to be swapped later. We will focus more on this issue when outlining the simulated annealing algorithm below.

5.7 Simulated Annealing

Algorithm 5.1 and the modified versions above only allow for edge-swaps that decrease the global cost function. That is, only edges that are not locally optimal can be swapped. The following example illustrates that "bad swaps", in the sense of swapping locally optimal edges such that the cost function increases, might be preferred as an intermediate step in a swapping strategy.

Assume that the triangulation in Figure 5.7(a) is the final (locally optimal) triangulation from the LOP with global cost $C_p(\Delta^a)$, calculated by (5.3), and that e_1 is an edge with a relatively high cost. Since e_1 is a diagonal in a non-convex quadrilateral it cannot be swapped. If, for example, the ABN cost function were used in the LOP, the high cost of e_1 may appear as a sharp edge in the triangulation with undesirable visual effects. Assume now that a "bad swap" of e_2 to e_2' is made producing the triangulation Δ^b in Figure 5.7(b). Since the triangulation in (a) was locally optimal, the cost function would increase by the swapping of e_2. But swapping e_2 makes e_1 swappable since it is now a diagonal in a convex quadrilateral. Swapping e_1 to e_1' may now result in a triangulation Δ^c with a global cost $C_p(\Delta^c) < C_p(\Delta^a)$ and which is preferable to the triangulation in Figure 5.7(a).

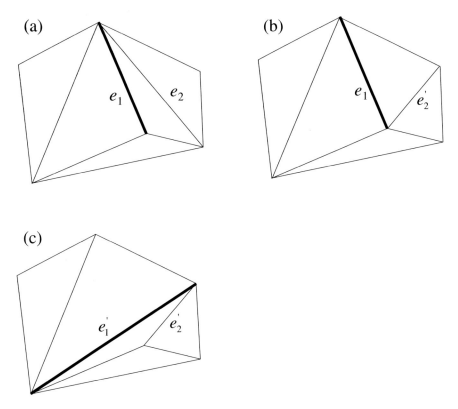

Fig. 5.7. A "bad swap" of e_2 followed by a swap of e_1 which reduces the clobal cost.

The observations from this simple exercise is the rationale for using *simulated annealing* for optimizing triangulations. Simulated annealing is a general method for solving large combinatorial minimization problems. A classical example is the problem of the traveling salesman who wants to find the shortest route when visiting a sequence of cities [46, 69].

Simulated annealing is also an analogy to processes in nature related to thermodynamics, for example how liquids freeze or how metals cool and anneal. The energy of a system is related to temperature by a probability distribution such that a system at low temperature is likely to have low energy. As the system cools the energy is likely to decrease, but there is always a probability that, at a certain temperature, the energy increases, though the chance for this gets smaller as the temperature decreases. Moreover, if a slow cooling process is applied, it is more likely that a global minimum energy state is reached than if the system is cooled down quickly. This is how nature finds its way towards a global minimum energy state avoiding getting stuck at local minima.

Simulated annealing schemes apply these principles in combinatorial optimization, and simulated annealing algorithms can be used for optimizing triangulations. Let the global cost function in (5.3) correspond to the energy of processes in nature. Further, let "bad swaps" that increase the global cost function correspond to nature's increase in energy which is more likely to occur at a high temperature than at a low temperature. The principle now is to choose edges from the triangulation randomly at each stage of the process. If the global cost function decreases by swapping an edge e_i, then e_i will be swapped as was the case in the LOP algorithm. But, if the global cost function is increased by swapping e_i, then a "bad swap" of e_i can also be made. The probability that a "bad swap" is made is reduced at each stage of the algorithm according to an *annealing schedule* defined as a sequence of decreasing real numbers, or "temperature" steps, $t_1 > t_2 > \cdots > t_{ntemps} > 0$. Let *nlimit* and *glimit* be positive integers. Further, let $d = C_p(\Delta') - C_p(\Delta)$ be the change in the global cost function (5.3) by swapping an edge. The simulated annealing Algorithm 5.2 was suggested by Baszenski & Schumaker [8] to construct optimal triangulations.

Algorithm 5.2 Simulated annealing

---

 1. **do** $k = 1, \ldots, ntemps$
 2. $t_k = r^k t_0, \quad 0 < r < 1$, e.g., $r = 0.95$
 3.   **do** $l = 1, \ldots, nlimit$
 4.   **while** the number of "good swaps" $\leq glimit$
 5.   let Δ be the current triangulation; choose a random edge e_i in Δ
 6.   **if** e_i is swappable
 7.   let Δ' be the result of swapping e_i and let $d = C_p(\Delta') - C_p(\Delta)$
              be the corresponding change of the global cost function (5.3).
 8.   **if** $d < 0$, i.e., if the global cost decreases
 9.   swap e_i ("good swap")
10.   **else**
11.   choose a random number θ, $0 \leq \theta \leq 1$
12.   **if** $\theta \leq e^{-d/t_k}$
13.   swap e_i ("bad swap")
14.   **endif**
15.   **endif**

The parameters *ntemps* and *nlimit* control the number of stages of the algorithm and the number swaps to be attempted at each stage respectively. The annealing schedule is defined by Step 2. With r close to unity, the annealing schedule is almost linear such that the temperature is reduced by a factor of r at each temperature step. Step 12, by the term e^{-d/t_k}, decides if a

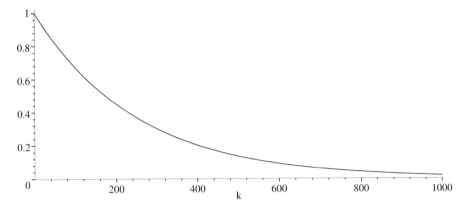

Fig. 5.8. Probability of making bad swaps in Step 12 of Algorithm 5.2.

"bad swap" should be made at a given temperature t_k, increasing the global cost by $d = C_p(\Delta') - C_p(\Delta)$. (The term e^{-d/t_k} corresponds to the Boltzmann probability distribution in thermodynamics.) If d is large, the probability of making a bad swap gets smaller, and the probability of making a bad swap is also reduced as one proceeds through the stages of the algorithm with decreasing temperatures t_k, see Figure 5.8. Note also the parameter *glimit* that controls the maximum number of good swaps allowed at each temperature step. Setting *glimit* too high corresponds to a system in nature that is cooled down too quickly. The consequence is that a minimum close to the global minimum is less likely to be found. The same effect can be observed if the initial temperature t_0 is set too low.

Compared to the LOP algorithm, simulated annealing is slow. It is also extremely sensitive to the choice of parameters. In practical applications, user interfaces must be provided to assist and guide the user, and to let the user experiment with different parameter values. A useful feature is to include a "polling function" in the inner loop of the algorithm such that the user can stop the process at any time and examine the result obtained so far. If the result is not satisfactory, the algorithm is reactivated with the current state of parameters and with the current result as the initial triangulation [4]. A thorough discussion of the choice of parameters and numerical experiments on optimizing triangulations can be found in [75].

Figure 5.9 shows the result of running Algorithm 5.2 starting from a Delaunay triangulation of the same data set as used previously. The level curves are not visually different from those in Figure 5.3 where the LOP algorithm was used. However, the triangles resulting from simulated annealing are even more elongated in directions where the magnitude of the second directional derivative of the underlying function is small. This indicates that a piecewise linear polynomial function defined over this triangulation gives a better

(a)   (b)

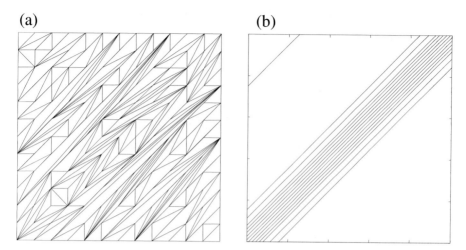

Fig. 5.9. Triangulation optimized with the simulated annealing algorithm using ABN criterion and l_1 norm.

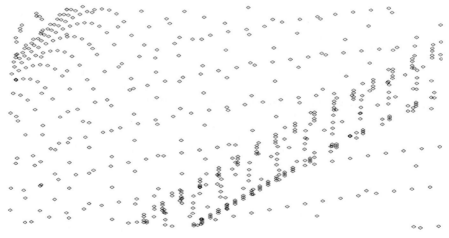

Fig. 5.10. 1108 points measured from an interior detail of a car.

approximation to the underlying function shown in Figure 5.1. Figure 5.13 shows another example of using simulated annealing, which can be compared with the result in Figure 5.12 from using the LOP algorithm. There are only minor differences in the level curves, but the triangles are longer and thinner, as observed in the example above.

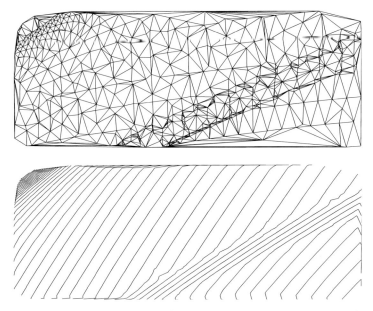

Fig. 5.11. Delaunay triangulation of the points in Figure 5.10, and level curves.

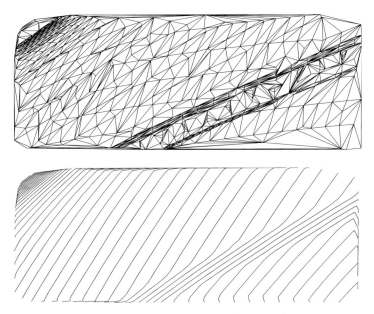

Fig. 5.12. Data dependent triangulation from LOP algorithm, SCO criterion and l_1 norm.

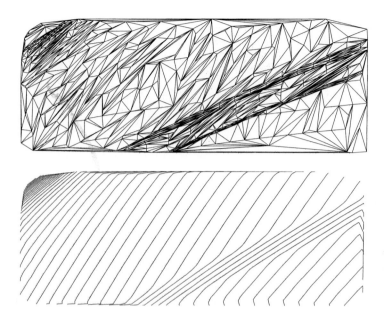

Fig. 5.13. Data dependent triangulation from the simulated annealing algorithm, SCO criterion and l_1 norm.

5.8 Exercises

1. Show that the vectors $\mathbf{n}^{(1)}$ and $\mathbf{n}^{(2)}$ in the ABN cost function have the same directions as the vectors $\mathbf{v}^{(1)}$ and $\mathbf{v}^{(2)}$ in the SCO cost function when $\mathbf{n}^{(1)}$ and $\mathbf{n}^{(2)}$ are projected onto a horizontal plane.
2. Compare the LOP algorithm 5.1 with the simulated annealing algorithm 5.2 and relate the optimization process to thermodynamic systems in nature. How is the "cooling process" of the LOP algorithm compared to that of simulated annealing?
3. Assume that a set of points is sampled from a surface that is smooth everywhere except for in certain regions where there are sharp edges. An initial Delaunay triangulation is constructed from the sampled data. Try to choose a cost function and a norm for optimization of the triangulation such that sharp edges are recovered and also such that the surface triangulation is smooth where the underlying surface is smooth. Choose from the cost functions and norms in this chapter, or combine them, or design your own. You may also consider modifying the optimization algorithms or making your own. Discuss. (Hint: If you decide to design your own local cost function, think of some discrete measure of Gaussian or mean curvature).

6

Constrained Delaunay Triangulation

The theory of Delaunay triangulation can be generalized to account for constrained edges also referred to as prespecified edges or break lines. This leads to the notion of *constrained Delaunay triangulation*[1] (CDT). Constrained edges may represent rivers, roads, lake boundaries and mountain ridges in cartography, or linear features in finite element grids. CDT may also be used to construct triangulations with holes and triangulations with arbitrarily shaped (non-convex) boundaries, while preserving Delaunay properties on the interior of the triangulation away from holes and boundaries.

This chapter extends and generalizes the theory of conventional Delaunay triangulation by including constrained edges. Algorithms for constructing CDTs are presented in detail. Chapters 3 and 4 should be read before this chapter. Alternatively, you can read the first two sections of this chapter even if you have not read Chapter 4.

6.1 Delaunay Triangulation of a Planar Straight-Line Graph

The notion of *constrained triangulation* was introduced in Section 1.4 together with a simple algorithm for its construction. In this chapter, we introduce the notion of *constrained Delaunay triangulation* (CDT), which is a generalization of the conventional Delaunay triangulation presented in Chapters 3 and 4. More specifically, we consider the triangulation of a set of points P and a set of edges E_c in the plane where P and E_c constitute a planar straight-line graph (PSLG), which we denote by $G(P, E_c)$ (cf. Section 2.1). We assume that the endpoints of edges in E_c are contained in P and it is required that E_c is a subset of the edges in the final triangulation $\Delta(G)$. The set E_c is said to be

---

[1] Some authors use the term *Generalized Delaunay triangulation*.

constrained edges of $\Delta(G)$. In addition, we will impose a Delaunay property on $\Delta(G)$, which will be made clear in what follows.

In the following, two points p_i and p_j are said to be *visible* to each other if the straight line segment between p_i and p_j does not intersect the interior of any edge in E_c or any untriangulated region, such as a hole or a region outside the exterior boundary of $\Delta(G)$. Recall Definition 3.3 of a Delaunay triangulation based on the circle criterion. By modifying the circle criterion, taking into account the constrained edges E_c, which in general are not Delaunay edges, we can now modify the definition of a conventional Delaunay triangulation to obtain the definition of a CDT.

Definition 6.1 (Constrained Delaunay triangulation, modified circle criterion). *A constrained Delaunay triangulation $\Delta(G)$ of a PSLG $G(P, E_c)$ is a triangulation containing the edges E_c such that the circumcircle of any triangle t in $\Delta(G)$ contains no point of P in its interior which is visible from all the three nodes of t.*

Hence, the circle criterion outlined in Section 3.5 can be violated such that points in P can be interior to circumcircles of triangles in $\Delta(G)$ if the points are "hidden" behind edges of E_c. We will refer to this as the *modified circle criterion* for CDT. The edges of $\Delta(G)$ that are not in E_c will be referred to as Delaunay edges in the CDT, and the triangles of $\Delta(G)$ are called Delaunay triangles. If E_c is an empty set, the graph $G(P, E_c)$ reduces to the point set P. Then the definition above becomes equivalent to Definition 3.3 such that the constrained Delaunay triangulation $\Delta(G)$ reduces to a conventional Delaunay triangulation $\Delta(P)$.

Figure 6.1 shows a PSLG $G(P, E_c)$ in (a), a Delaunay triangulation of P in (b) and a CDT of $G(P, E_c)$ in (c). In (d) we see a circumcircle of a triangle that contains two points of P in its interior, but the points are "hidden" behind a constrained edge and thus not visible from all the three nodes of the triangle which means that the modified circle criterion holds for that triangle.

In Chapter 3 we also gave two other equivalent characterizations of a Delaunay triangulation: the Delaunay triangulation of a set of points P was defined to be the optimal triangulation among all possible triangulations of P, in the sense of the MaxMin angle criterion (Definition 3.1). This definition also applies directly to the constrained case when considering all possible triangulations of a PSLG $G(P, E_c)$, where each "possible triangulation" of G is constrained to contain all the edges in E_c. Definition 3.2 of a conventional Delaunay triangulation was based on the Voronoi diagram; the Delaunay triangulation was defined as the straight-line dual of the Voronoi diagram. This is not immediately transferable to the constrained case. Joe & Wang [45] regard constrained edges as obstacles and define a *constrained Voronoi diagram* of a PSLG by modifying the distance metric such that $d(p_i, p_j) = \infty$ if the points p_i and p_j are not visible to each other. An extension of the constrained

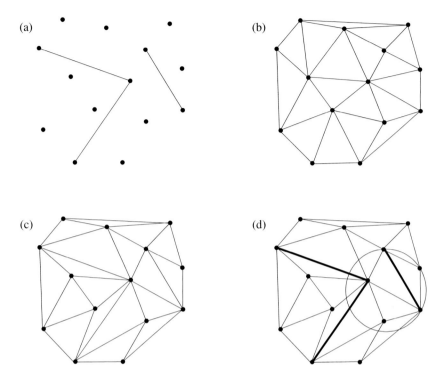

Fig. 6.1. (a): A planar straight-line graph $G(P, E_c)$. (b): Conventional Delaunay triangulation of the point set P. (c): Constrained Delaunay triangulation of $G(P, E_c)$. (d): Illustration of the modified circle criterion for constrained Delaunay triangulation.

Voronoi diagram to a larger space that contains R^2 is then shown to be the dual of the CDT. This theory is quite involved and will not be treated in any depth in this chapter.

6.2 Generalization of Delaunay Triangulation

In this section we generalize the theory of conventional Delaunay triangulation to obtain a theory for constrained Delaunay triangulation of a planar straight line graph $G(P, E_c)$. The approach will be somewhat different to that of Chapter 3 since we do not rely on a dual construction, a Voronoi diagram, which was the basis for the final results of the theory for conventional Delaunay triangulation. Proofs of lemmas and theorems are not given in detail here as most of them will be quite similar to the proofs in Chapter 3. The paper by Lee & Lin [51] can be consulted for details of proofs and completeness of

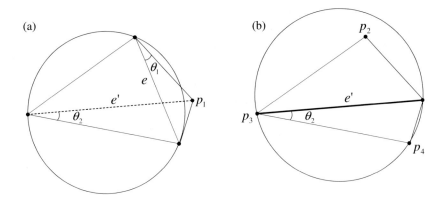

Fig. 6.2. The triangulation in (a) with e as an edge is a conventional Delaunay triangulation. The triangulation in (b) is a constrained Delaunay triangulation due to the constrained edge e'.

the theory. To simplify the presentation, we assume, as before, that no four points in the point set P are cocircular.

First we state the equivalence of two of the Delaunay swapping criteria for strictly convex quadrilaterals (cf. Lemma 3.2 in Section 3.6).

Lemma 6.1. *The modified circle criterion and the MaxMin angle criterion are equivalent for strictly convex quadrilaterals.*

In the unconstrained case, the minimum interior angle of the two possible triangulations of a strictly convex quadrilateral is maximized if the diagonal is chosen such that the circle criterion holds. This is illustrated by choosing the edge e instead of e' in Figure 6.2(a). The angle θ_1 is strictly larger than θ_2 since the node p_1 is strictly outside the drawn circumcircle. In the constrained case we must also consider constrained edges, which are always chosen as diagonals of strictly convex quadrilaterals. If the edge e' is constrained, then the triangulation in Figure 6.2(b) is a CDT although the smallest interior angle θ_2 is not maximized. Also, the node p_2 is inside the drawn circumcircle, but since p_2 is not visible from p_4, the modified circle criterion is not violated and the triangulation is a CDT according to Definition 6.1.

Recall from Section 3.1 that an indicator vector was constructed for a triangulation such that each triangle in the triangulation has a corresponding entry in the indicator vector, which represents the smallest interior angle of that triangle. The entries in the vector are sorted in non-decreasing order and two indicator vectors corresponding to two different triangulations can be compared by their lexicographical measure. The following lemma, which is equivalent to Lemma 3.4 for a conventional Delaunay triangulation, can now be established for a CDT.

Lemma 6.2. *The indicator vector becomes lexicographically larger each time an edge of a strictly convex quadrilateral is swapped according to the Delaunay swapping criteria.*

This result is the basis for Lawson's local optimization procedure LOP (Algorithm 3.2) to construct a Delaunay triangulation from a set of points in the plane. In the following, we apply the LOP to a triangulation of a PSLG $G(P, E_c)$. Starting with an arbitrary triangulation of $G(P, E_c)$, the edges of strictly convex quadrilaterals are examined and swapped if they do not meet the constrained Delaunay criteria (cf. Lemma 6.1). This process is repeated until no edge-swap occurs. Due to Lemma 6.2 above, and since the number of possible triangulations of $G(P, E_c)$ is finite, the LOP terminates after a finite number of edge-swaps.

In the unconstrained case, it was shown that the LOP yields a triangulation that is Delaunay and that the lexicographical measure of the indicator is maximized. In the analysis that follows, we demonstrate that this also holds when the LOP is applied to an arbitrary constrained triangulation of a PSLG.

Recall that an edge in a triangulation is called *locally optimal* if the decision is not to swap it when applying the local optimization procedure LOP (Section 3.8). For a CDT we define the constrained edges as locally optimal by default, and the boundary edges of the triangulation are also locally optimal. Thus, after the LOP applied to a constrained triangulation has converged, all the edges are locally optimal. This also leads to the following global property of the triangulation similar to that stated in Theorem 3.2 in the unconstrained case.

Theorem 6.1. *All interior edges of a triangulation $\Delta(G)$ of a PSLG $G(P, E_c)$ are locally optimal if and only if each triangle of $\Delta(G)$ satisfies the modified circle criterion, i.e., if and only if the circumcircle of any triangle t in $\Delta(G)$ contains no point of P in its interior which is visible from all the three nodes of t.*

Proof. See Exercise 1.   □

Since all edges of a PSLG are locally optimal after the LOP has converged, the theorem implies that the LOP yields a constrained Delaunay triangulation in accordance with Definition 6.1. Similar to the unconstrained case we also find that a CDT, which is locally optimal, is also globally optimal.

Theorem 6.2. *A triangulation of a PSLG is a constrained Delaunay triangulation (in agreement with Definition 6.1) if and only if its indicator vector is lexicographically maximum.*

This result corresponds to Theorem 3.5 in the unconstrained case where the converse part of the proof was based on the *uniqueness* of a Delaunay triangulation. However, uniqueness was based on the dual construction, the

Voronoi diagram, which is not the basis for the theory in this chapter. The converse part of the proof must therefore follow another line, which is quite involved [51] (Exercise 6). However, the important property of uniqueness can also be proved for a CDT: if no four points of P in a PSLG $G(P, E_c)$ are cocircular, the constrained Delaunay triangulation $\Delta(G)$ is unique. We also leave this proof to the reader in Exercise 6.

The following theorem, which is a generalization of Theorem 3.1, gives a unique characterization of a Delaunay edge in a CDT[2].

Theorem 6.3. *For any PSLG $G(P, E_c)$, an edge e_{ij} between two points p_i and p_j of P is a Delaunay edge in the CDT of $G(P, E_c)$ if and only if p_i and p_j are visible to each other and there exists a circle passing through p_i and p_j that does not contain any points of P in its interior visible from both p_i and p_j.*

Proof. See Exercise 2.   □

Finally we give the generalization of Theorem 4.1 which uniquely defines a triangle in a CDT.

Theorem 6.4. *A triangle t with nodes p_i, p_j and p_k of a PSLG $G(P, E_c)$ is a triangle in the CDT $\Delta(G)$ if and only if the circumcircle of t contains no point of P in its interior which is visible from both p_i, p_j and p_k.*

Proof. See Exercise 3.   □

6.3 Algorithms for Constrained Delaunay Triangulation

Constrained Delaunay triangulations can be computed by similar schemes to those in Chapter 3 for conventional Delaunay triangulations. In fact, some algorithms need only be modified slightly by taking into account the constrained edges and using a different Delaunay criterion, the modified circle criterion in Definition 6.1. In this chapter, we only consider incremental algorithms for computing a CDT. Non-incremental algorithms for computing a CDT do exist, for example divide-and-conquer algorithms and other static algorithms that treat all constraints simultaneously [51], but this will not be considered here as the incremental approach leads to simpler and more flexible concepts and algorithms, as was the case with algorithms for conventional Delaunay triangulation in Chapter 4. We simplify the task of computing a CDT into two basic operations:

- inserting a constrained edge into an existing CDT, and
- inserting a new node into an existing CDT.

---

[2] Recall from Section 6.1 that the edges in $\mathbf{E}_c$ are not referred to as Delaunay edges.

With these two operations available, one can compute a CDT of a PSLG $G(P, E_c)$ incrementally. One can, for example, start with a conventional Delaunay triangulation of the point set P, computed with one of the algorithms in Chapter 4, and then insert one constrained edge from E_c at a time until the final CDT $\Delta(G)$ is obtained. We assume that the constrained edges do not intersect. If the constrained edge falls on some existing node, one may split the edge in that node and treat the constraint as two separate constrained edges. With the second operation above, one can also add nodes into an existing CDT.

After each point or edge insertion, the triangulation is updated such that the modified circle criterion holds for all triangles. Thus, the triangulation will always agree with Definition 6.1 of a constrained Delaunay triangulation. Recall from Section 4.4 that there are conceptually two ways to update a triangulation to be Delaunay after the insertion of a point:

- remove triangles that are not Delaunay and retriangulate the affected region,
- split the triangle which contains the insertion point into three new triangles and apply a recursive swapping procedure.

The same principles can be used when inserting a point or a constrained edge into an existing CDT. The rest of this chapter is devoted to procedures for constructing a CDT using these principles.

6.4 Inserting an Edge into a CDT

Let e_c be a constrained edge that is inserted between two existing nodes p_a and p_b in an existing constrained Delaunay triangulation $\Delta(P, E_c)$, see Figure 6.3(a). The constrained edge intersects a set of edges in their interior, and these edges must be removed from the triangulation. Since no new nodes are added by the insertion of e_c, it follows readily from Theorem 6.3 above that all edges that are not intersected by e_c should be preserved as edges in the new CDT, which we denote by $\Delta(P, E_c \cup e_c)$. Also, triangles that are not intersected in their interior by e_c still satisfy the modified circle criterion by the presence of e_c, and should therefore be preserved as triangles in $\Delta(P, E_c \cup e_c)$.

Thus, the *influence region of e_c in $\Delta(P, E_c)$* (cf. Section 4.5), which we denote by R^{e_c}, is a connected region composed of the union of only those triangles that are intersected in their interior by e_c. The constrained edge splits R^{e_c} into two connected regions with e_c as a common edge. Let $Q^{e_c,L}$ and $Q^{e_c,R}$ be the boundaries of these two regions, each containing e_c, obtained by removing only those edges that are intersected in their interior by e_c. $Q^{e_c,L}$ and $Q^{e_c,R}$ are to the left and to the right respectively of the directed line from p_a to p_b. We observe that $Q^{e_c,L}$ and $Q^{e_c,R}$ are not necessarily simple polygons. For example, $Q^{e_c,L}$ in Figure 6.3(b) contains an edge with an endpoint that

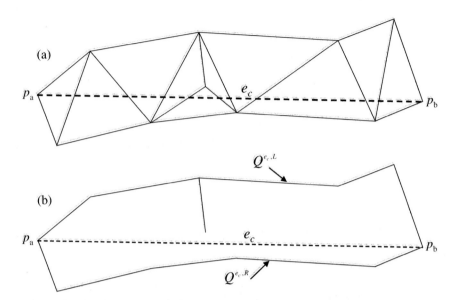

Fig. 6.3. (a) The influence region of a constrained edge e_c that is inserted into an existing triangulation. (b) the influence polygons $Q^{e_c,L}$ and $Q^{e_c,R}$ of e_c.

is not shared by another edge in $Q^{e_c,L}$. This makes no difference for the triangulation procedures described in the following. We call $Q^{e_c,L}$ and $Q^{e_c,R}$ the *influence polygons of e_c in* $\Delta(P, E_c)$.

Provided that the CDT $\Delta(P, E_c \cup e_c)$ exists, it is evident that $\Delta(P, E_c \cup e_c)$ is obtained by triangulating the influence polygons $Q^{e_c,L}$ and $Q^{e_c,R}$ on each side of the constrained edge e_c separately while ensuring that the modified circle criterion holds for all new triangles. Recall the step-by-step approach for making Delaunay triangles from a set of points P in Section 4.3. The procedure was based on the principle of a circle growing from an existing edge, a *base line* e_b, until a new point p was reached to form a new triangle in the triangulation. This process was applied recursively with a new base line e_b each time, where e_b was an edge created in the previous step. The growing circle always interpolates the endpoints of e_b, and p is chosen as the first point in P reached by the growing circle, or equivalently, as the point p in P that makes the largest angle at p spanned by the base line. The circle is point-free initially, and it remains point-free when it grows to reach a new point. Further, each new growing circle in the recursion starts with a point-free circle from the previous step in the recursion. Thus, no point from P ever falls strictly inside the growing circle, and this ensures that the (conventional) circle criterion holds for the new triangles.

Only a slight modification of this procedure is needed to retriangulate the influence polygons $Q^{e_c,L}$ and $Q^{e_c,R}$ of the constrained edge e_c. Assume that

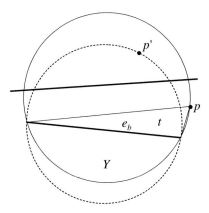

Fig. 6.4. The growing circle reaches a point p' first, but p' is separated from e_b by a constrained edge and cannot form a triangle with e_b.

the first base line e_b is the constrained edge e_c. The growing circle interpolates the endpoints of e_c and is infinitely large initially with center on the opposite side of e_c from the influence polygon to be triangulated. The first point reached by a growing circle may be separated from the base line by an existing constrained edge. This is illustrated in Figure 6.4. The dotted circle has reached the point p' on the opposite side of a constrained edge from e_b, but a triangle cannot be formed by p' and the endpoints of e_b since the new edges would intersect the constrained edge. On the other hand, the next point p reached by the growing circle defines a legal triangle t with e_b. We also note that any point inside the circle disk Y on the opposite side of e_b from p is not visible from p; thus, the modified circle criterion now holds for t if e_b is a constrained edge. Since t is the only possible new triangle that can be constructed from e_b, we conclude that t is Delaunay.

The point p that defines the new triangle with the base line e_b is now uniquely defined as the point in P that,

i) makes the largest angle at p spanned by the base line e_b, and
ii) p is visible from the endpoints of e_b.

This triangle construction is done recursively with a new base line e_b each time, as explained above for conventional Delaunay triangulation, until the influence polygon is covered by triangles. Figure 6.5(a)–(c) illustrates the recursion for the upper influence polygon $Q^{e_c,L}$ of e_c, and (d) shows the final retriangulation of the whole influence region R^{e_c}.

The modified circle criterion is guaranteed to hold for all new triangles. Any point from P that falls strictly inside the growing circle when the circle has reached a point p, will always be separated from p by e_c or one or more existing constrained edges and thus not visible from p which becomes a node in the new triangle.

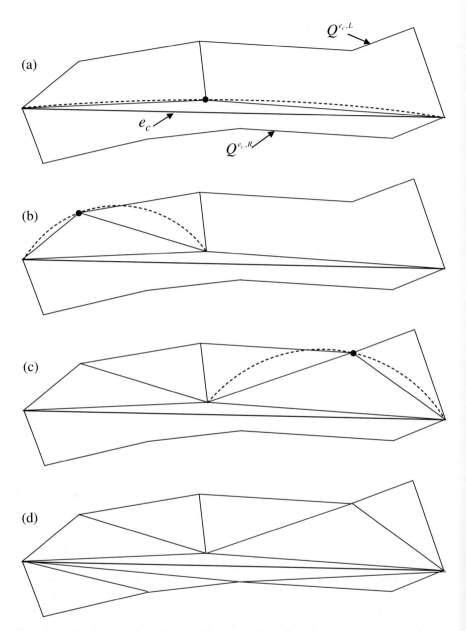

Fig. 6.5. Retriangulation of the influence region of a constrained edge e_c. (a) to (c) show how new triangles are constructed when retriangulating the left influence polygon $Q^{e_c,L}$. The growing circles (dotted) are shown when they have reached a point where a new triangle can be formed with the base line. In (d) the constrained Delaunay triangulation of the whole influence region is shown.

6.5 Edge Insertion and Swapping

When inserting a constrained edge e_c between two existing nodes in a triangulation Δ, there are no new nodes added to Δ, and consequently, from Equation (1.2) and (1.3), the number of triangles and edges is not changed. If the boundary of Δ were convex, then it follows from Theorem 3.4 that, at least theoretically, the new triangulation containing e_c can be reached by a sequence of edge-swaps starting from Δ. Dyn & Goren & Rippa [23] have shown that it is possible to transform any triangulation of a set of points inside an arbitrary polygonal domain into any other triangulation of the same set of points using only operations of edge-swapping. In this section, we present the necessary theoretical results from their work, which prepares the ground for a swapping algorithm for inserting a constrained edge e_c into an existing CDT. The polygonal domain will now be the influence region R^{e_c} of e_c in Δ, that is, the union of the triangles intersected by e_c. The algorithm has two main steps:

1. A swapping procedure for including e_c as an edge in the existing triangulation. Only edges intersected by e_c will be swapped. This will produce a new triangulation Δ' that is not necessarily a CDT, but it contains e_c as an edge.
2. Lawson's local optimization procedure LOP is then applied to edges inside each of the two influence polygons $Q^{e_c,L}$ and $Q^{e_c,R}$ to obtain the CDT with e_c as a constrained edge.

So, how can we accomplish Step 1 above by swapping edges inside R^{e_c} such that eventually e_c is incorporated as an edge in the triangulation? Before answering this question we need some theoretical results.

Consider again the example of the previous section, illustrated in Figure 6.6, where an edge e_c is inserted between two existing nodes p_a and p_b. Let the ordered set of vertices $(p_a, u_1, \ldots, u_n, p_b)$ define the closed influence polygon $Q^{e_c,L}$ to the left of the directed line from p_a to p_b. Note how the vertices of the polygon are enumerated when it is not simply connected as in Figure 6.6(a). In the presence of an edge in $Q^{e_c,L}$ that has an endpoint that is not shared by another edge in $Q^{e_c,L}$, a subset of the sequence of the u_is defines a loop, for example (u_2, u_3, u_4) in the figure. Thus, one or more pairs (u_l, u_k) may refer to the same vertex of $Q^{e_c,L}$ as (u_2, u_4) in the figure. We say that such polygons are multiply connected. A unique interior angle α_i of the closed polygon $Q^{e_c,L}$ corresponds to each u_i.

The principle of the swapping procedure to be detailed below is to swap edges away from e_c such that eventually e_c is included as a constraind edge in the triangulation. More specifically, any edge that is swapped intersects with e_c and has its endpoints among the vertices of the influence polygons $Q^{e_c,L}$ and $Q^{e_c,R}$ on each side of e_c. Further, any swappable edge must be a

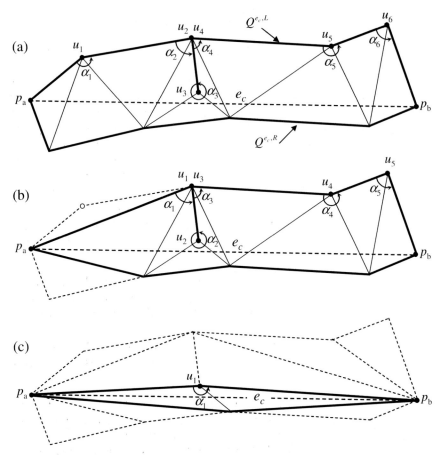

Fig. 6.6. Illustration for edge insertion and swapping. (b) shows the situation when the point u_1 has been isolated from the influence polygon $Q^{e_c,L}$, and (c) shows the situation when the last edge is swapped and takes on the role as the constrained edge e_c.

diagonal in a convex quadrilateral, which implies that the interior angles at the vertices of $Q^{e_c,L}$ and $Q^{e_c,R}$ corresponding to the endpoints of the swapped edge must be smaller than π.

In the following, we prove that there always exists a swappable edge among the edges, that intersects with e_c. We first show that it is possible to find a vertex u_m of $Q^{e_c,L}$ where the interior angle $\alpha_m < \pi$, and next we show that at least one of the edges radiating from u_m and intersecting e_c has its other endpoint at a vertex of $Q^{e_c,R}$ where the interior angle is also smaller than π. Thus, there is at least one edge intersecting e_c which is a diagonal of a convex quadrilateral and thus swappable. These observations lead to an elegant swapping procedure for including a constrained edge in a triangulation.

Lemma 6.3. *A closed and simply connected polygon $\mathcal{P}$ has at least three interior angles smaller than π.*

Proof. Let $\alpha_1, \ldots, \alpha_N$, $N \geq 3$, be the interior angles of $\mathcal{P}$ and suppose that k of them are smaller than π. From elementary geometry we know that the sum of all interior angles in a closed polygon is $(N-2)\pi$. We partition this sum in angles smaller than π and angles greater or equal to π,

$$(N-2)\pi = \sum_{i=1}^{N} \alpha_i = \sum_{\alpha_i < \pi} \alpha_i + \sum_{\alpha_i \geq \pi} \alpha_i. \tag{6.1}$$

Since k of the N angles are smaller than π, the sum of those greater or equal to π satisfy

$$\sum_{\alpha_i \geq \pi} \alpha_i \geq (N-k)\pi.$$

Replacing the last term on the right-hand side in (6.1) with $(N-k)\pi$, we then get the bound

$$(N-2)\pi \geq \sum_{\alpha_i < \pi} \alpha_i + (N-k)\pi,$$

and reordering we get

$$(k-2)\pi \geq \sum_{\alpha_i < \pi} \alpha_i.$$

The right-hand side, which expresses the sum of the k angles smaller than π, is non-negative, so $k \geq 2$ due to the inequality. But since $k > 0$, the right-hand side is in fact positive. And since the right-hand side is positive, the left-hand side is also positive due to the inequality, and therefore $k \geq 3$. □

For a simply connected influence polygon, Lemma 6.3 can be applied directly to conclude that there is at least one interior angle different from those at the endpoints of e_c that is smaller than π. It can also be shown that the lemma holds for a multiply connected polygon as $Q^{e_c,L}$ in Figure 6.6(a). (Exercise 4.)

So, using the notation above, there is at least one point u_m where the interior angle $\alpha_m < \pi$. (Consult Figure 6.7.) Next we show that at least one of the edges radiating from u_m and intersecting e_c is a diagonal in a convex quadrilateral and thus swappable. Suppose that there are r edges radiating from u_m that intersect the interior of e_c. The endpoints of the r edges on the opposite side of e_c from u_m are denoted by $w_1, \ldots, w_r$, numbered as in the figure. We also use the conventions $w_0 = u_{m-1}$ and $w_{r+1} = u_{m+1}$.

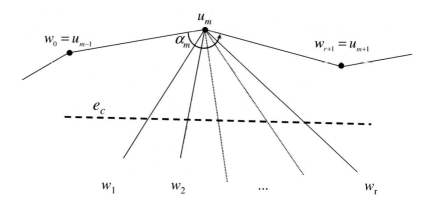

Fig. 6.7. Illustration for Lemma 6.4.

Lemma 6.4. *Let u_m be a point on the influence polygon where the interior angle $\alpha_m < \pi$. Then there is at least one edge radiating from u_m and intersecting the interior of e_c that is a diagonal in a convex quadrilateral.*

Proof. The closed polygon defined by the sequence $(w_0, w_1, \ldots, w_r, w_{r+1})$ has at least three interior angles smaller than π by Lemma 6.3. (Note that the sequence does not include u_m.) Thus, there is at least one point w_s, $1 \le s \le r$ such that the angle $\angle w_{s-1} w_s w_{s+1}$ is smaller than π. Then the quadrilateral with (u_m, w_s) as a diagonal must be convex since $\angle w_{s+1}, u_m, w_{s-1}$ is also smaller than π.   □

This result ensures that all edges (u_m, w_s), $1 \le s \le r$, where $\alpha_m < \pi$, can be swapped away from u_m such that there are no edges left radiating from u_m and intersecting the interior of e_c. This can be done to eliminate u_m from the influence polygon $Q^{e_c, L}$. In Algorithm 6.1, r is still the number of edges radiating from u_m and intersecting the interior of e_c.

Algorithm 6.1 Eliminate u_m, with $\alpha_m < \pi$, from $Q^{e_c, L}$

---
1. **while** $(r \ge 1)$
2.   Let (u_m, w_s) be a diagonal in a convex quadrilateral $(u_m, w_{s-1}, w_s, w_{s+1})$.
3.   Swap (u_m, w_s) to (w_{s-1}, w_{s+1})
4. $r \leftarrow r - 1$

When $r = 1$ there is only one edge (u_m, w_1) left radiating from u_m and intersecting e_c. This edge is a diagonal in the quadrilateral $(u_m, u_{m-1}, w_1, u_{m+1})$ that is also convex by Lemma 6.4. When (u_m, w_1) is swapped to (u_{m-1}, u_{m+1}) in the r'th cycle of the algorithm, u_m is isolated from e_c by the last swapped edge (u_{m-1}, u_{m+1}). We can now eliminate u_m from $Q^{e_c,L}$, and obtain a new influence polygon defined by the sequence $(p_a, u_1 \ldots, u_{m-1}, u_{m+1}, \ldots u_n, p_b)$. Figure 6.6(b) shows an example where the point u_1 from 6.6(a) has been eliminated from $Q^{e_c,L}$ through two edge-swaps. We also observe that the influence polygon $Q^{e_c,R}$ on the other side of e_c has been reduced by one point.

Algorithm 6.1 can be applied repeatedly with a new point u_m each time, until all points u_i, $i = 1, \ldots, n$ are eliminated from the original influence polygon $Q^{e_c,L} = (p_a, u_1, \ldots, u_n, p_b)$. Algorithm 6.2 illustrates the process.

Algorithm 6.2 Include e_c as an edge in a triangulation

---

1. **while** $(n >= 1)$
2. Find a point u_m, $1 \leq m \leq n$ where $\alpha_m < \pi$
3. Apply Algorithm 6.1 to u_m
4. $n \leftarrow n - 1$
5. $Q^{e_c,L} \leftarrow (p_a, u_1, \ldots, u_{m-1}, u_{m+1}, \ldots, u_n, p_b)$

Lemma 6.3, extended to hold for multiply connected closed polygons (Exercise 4), guarantees that a point u_m is always found in Step 2. When $n = 1$, the influence polygon is defined by (p_a, u_1, p_b) and the interior angle α_1 at u_1 is smaller than π by Lemma 6.3. In the last cycle of Algorithm 6.1, when it is applied to the closed polygon (p_a, u_1, p_b), the convex quadrilateral is defined by (u_1, p_a, w_1, p_b) with (u_1, w_1) as a diagonal, see Figure 6.6(c). This is the only edge left intersecting e_c, and when (u_1, w_1) is swapped to (p_a, p_b), it takes on the role as the constrained edge e_c that has p_a and p_b as endpoints.

Finally, the local optimization procedure is applied to edges inside the original influence polygons $Q^{e_c,L}$ and $Q^{e_c,R}$, which have been swapped away from e_c. This process converges (see Section 6.2) and the result is a CDT with e_c as a constrained edge.

In Section 3.4, we stated the existence of a conventional Delaunay triangulation. From the results above, it is evident that a constrained edge e_c can always be incorporated into a conventional Delaunay triangulation or into an existing CDT, $\Delta(P, E_c)$, to obtain a new modified CDT, $\Delta(P, E_c \cup e_c)$, provided that e_c does not intersect any existing constrained edges in E_c. Thus, the property of existence can also be stated for a CDT.

6.6 Inserting a Point into a CDT

Let $\Delta(P \cup p, E_c)$ be the CDT obtained by inserting a point p into a CDT $\Delta(P, E_c)$. As in the unconstrained case in Section 4.5 we call the region R^p that needs to be modified by the insertion of p the *influence region* of p in $\Delta(P, E_c)$. The external boundary of R^p, denoted Q^p, is called the *influence*

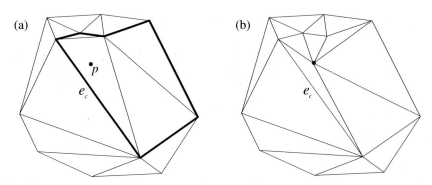

Fig. 6.8. (a): The influence polygon Q^p of a point p in a CDT $\Delta(P, e_c)$ is shown with bold edges. (b): The updated CDT $\Delta(P \cup p, e_c)$.

polygon of p in $\Delta(P, E_c)$. Generalization of Lemma 4.1 and Theorem 4.2 gives an exact limitation of R^p and shows how the new triangulation can be obtained when inserting p.

Lemma 6.5. *A triangle t in $\Delta(P, E_c)$ will be modified when inserting a point p to obtain $\Delta(P \cup p, E_c)$ if and only if the circumcircle of t contains p in its interior and p is visible from all the three nodes of t.*

Proof. The proof follows directly from Theorem 6.4. □

Figure 6.8(a) shows the influence polygon of a point p that is inserted into an existing CDT with a constrained edge e_c.

Theorem 6.5. *All new triangles of $\Delta(P \cup p, E_c)$ have p as a common node.*

Proof. See Exercise 5. □

Exactly as in the unconstrained case, it follows that $\Delta(P \cup p, E_c)$ can be obtained from $\Delta(P, E_c)$ by removing all triangles of R^p and connecting p to all points of Q^p, see Figure 6.8(b).

In Section 4.6, we developed a swapping procedure for inserting a point p into a conventional Delaunay triangulation. First, the triangle t containing p was located and t was split into three new triangles with p as a common node. Then, the recursive swapping procedure, Algorithm 4.3, was applied to the three edges e_1, e_2 and e_3 of t. We proved that the resulting triangulation on termination of the algorithm was Delaunay. This procedure only needs a slight modification to be valid for a CDT. The circumcircle test acting on an edge e_i must be preceded by checking if the edge is among the constrained edges E_c, in which case e_i should not be swapped and the recursion branch broken. We get Algorithm 6.3 to be applied to e_1, e_2 and e_3. The algorithm differs from its unconstrained counterpart only in the two first steps. The notation refers to Figure 4.5.

Algorithm 6.3 `recSwapDelaunayConstr(Edge $e_i$)`

1. **if** $(e_i \in E_c)$
2.   **return**
3. **if** `(circumcircleTest($e_i$) == true)`   // Algorithm 3.1 in Section 3.7
4.   **return**
5. `swapEdge($e_i$)`   // the swapped edge e_i' is incident with p
6. `recSwapDelaunay($e_{i,1}$)`   // call this procedure recursively
7. `recSwapDelaunay($e_{i,2}$)`   // call this procedure recursively

A useful operation for later use is to insert a new node on an existing constrained edge, thus splitting it into two new contiguous edges that are also constrained. In Chapter 7 this operation is used in an algorithm for generating meshes for the finite element method (FEM). The problem is left to the reader in Exercise 7.

6.7 Exercises

1. Prove Lemma 6.1 by adjusting the proof of Theorem 3.2 to fit a CDT.
2. Prove Theorem 6.3 which is a generalization of Theorem 3.1 in the unconstrained case. Note that the proof cannot rely on a Voronoi diagram.
3. Prove Theorem 6.4.
4. Show that Lemma 6.3 also holds for a multiply connected closed polygon.
5. Prove Theorem 6.5. Note that the proof cannot rely on a Voronoi diagram.
6. Prove Theorem 6.2. Hint: for the converse part of the proof, i.e., for proving that a CDT $\Delta(G)$ must have an indicator vector that is lexicographically maximum:
   a) First read the remark after Theorem 6.2.
   b) Let $\Delta(G)$ be the constrained Delaunay triangulation and use Theorem 6.1 to characterize it.
   c) Let $\Delta'(G)$ be a CDT that *is* lexicographically maximum, but $\Delta'(G) \neq \Delta(G)$. Then there must exist a Delaunay edge in $\Delta'(G)$ that intersects an edge of $\Delta(G)$. Discuss this situation and conclude that $\Delta(G)$ must be lexicographically maximum. (Remember to take constrained edges into account).
   d) Try also to deduce from the proof that a CDT is unique (with the common assumption that no four points are cocircular).
7. Derive a swapping procedure for splitting a constrained edge in a CDT into two adjacent constrained edges. That is, a point should be inserted on an existing constrained edge. The final triangulation should be a CDT.

7

Delaunay Refinement Mesh Generation

Refinement of triangulations is motivated by grid generation for the finite element method (FEM). Having said this, it is important to note that these techniques are also applicable in visualization, scattered data interpolation and other applications demanding triangulations with well-shaped triangles. The refinement scheme presented in this chapter successively refines a constrained triangulation by node insertion until the triangulation satisfies certain criteria on the shape of triangles. Roughly speaking, we aim at creating triangulations with triangles as equilateral as possible while simultaneously satisfying given constraints. As the title indicates, Delaunay properties will be maintained throughout the refinement process.

7.1 Introduction

Finite element methods (FEM) use triangulations as a basis for solving partial differential equations numerically. The process of creating triangulations (and other types of meshes) for FEM is called *gridding* or *meshing*. When operating in two dimensions, a planar domain is subdivided into a finite element mesh of elements. Differential equations, for example representing heat distribution or airflow, are then approximated with piecewise polynomial functions over the finite element mesh and solved numerically to achieve an approximate solution of the original problem.

In this chapter, we will be concerned with meshing techniques that generate triangulations only, and further, we will only consider meshing techniques that preserve Delaunay properties in the plane. This is also the rationale for the title of the chapter, which is borrowed from Shewchuk [79]. *Refinement* in the title refers to the process of how a mesh is generated: starting with a coarse mesh, which is a constrained Delaunay triangulation (CDT) of a planar straight line graph (PSLG), the mesh is successively refined by inserting new nodes into the CDT similar to incremental procedures for Delaunay triangulation treated in detail in Chapters 4 and 6. After each node insertion, the triangulation is updated to be Delaunay or constrained Delunay. The main difference from the scattered data interpolation problem that motivated us earlier is that only part of the data, a PSLG, is given in advance while insertion of new nodes into the mesh follows a predetermined meshing scheme. The initial PSLG can, for example, be an exterior boundary, possibly non-convex, together with isolated nodes and linear features inside the boundary, some of which may be closed polygons representing holes in the geometry. See the upper leftmost graph of Figure 7.1 for an example.

In this chapter, we examine a scheme that originates from Chew [14, 15] and Ruppert [73]. Shewchuk has unified, extended and refined the works of Chew and Rupert in [79, 80], and presents an implementation of Ruppert's scheme in [77]. These publications are the main sources for the presentation that follows. More details and advice on implementation can be found in Shewchuk's publications.

7.2 General Requirements for Meshes

The running time for solving systems of equations that arise from approximate solutions of partial differential equations over FEM meshes depends on the number of mesh elements. On the other hand, the quality of the approximate FEM solution, that is, how well it approximates the exact solution, depends on the granularity of the FEM mesh; a finer mesh with many small triangles generally gives a more accurate solution than a coarse mesh with fewer triangles. So there are two opposing demands on mesh density; it should be minimal with regards to running time but sufficiently dense to ensure accuracy of the approximate solution. An adaptive mesh with varying density over the domain makes the compromise easier. An adaptive mesh may be justified by the fact that in many applications an accurate solution is needed only in certain regions of the domain, while less accuracy suffices in regions of less importance. Moreover, in many applications, high mesh density is needed only in regions where the solution to the FEM problem varies rapidly. So, an adaptive mesh should aim at varying the triangle size over the domain with more and smaller triangles in regions with high accuracy demands, and we aim for a nice *spatial grading* from small triangles to larger triangles in areas of less importance.

Furthermore, the accuracy of the approximate solution, the numerical stability and the convergence are also affected by the *shape* of the triangles. Elongated triangles with small or large interior angles may lead to unstable computations. This is also a problem in surface modeling with polynomial functions defined over triangles. In the latter case, poorly shaped triangles may cause oscillations and visual artifacts caused by large variations in the

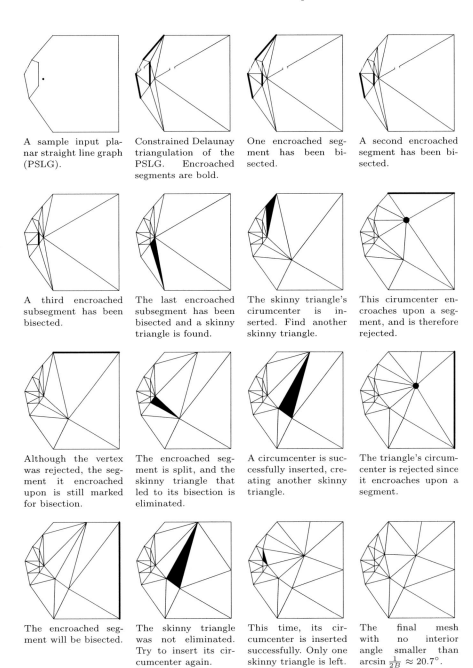

A sample input planar straight line graph (PSLG).

Constrained Delaunay triangulation of the PSLG. Encroached segments are bold.

One encroached segment has been bisected.

A second encroached segment has been bisected.

A third encroached subsegment has been bisected.

The last encroached subsegment has been bisected and a skinny triangle is found.

The skinny triangle's cirumcenter is inserted. Find another skinny triangle.

This cirumcenter encroaches upon a segment, and is therefore rejected.

Although the vertex was rejected, the segment it encroached upon is still marked for bisection.

The encroached segment is split, and the skinny triangle that led to its bisection is eliminated.

A circumcenter is successfully inserted, creating another skinny triangle.

The triangle's circumcenter is rejected since it encroaches upon a segment.

The encroached segment will be bisected.

The skinny triangle was not eliminated. Try to insert its circumcenter again.

This time, its circumcenter is inserted successfully. Only one skinny triangle is left.

The final mesh with no interior angle smaller than $\arcsin \frac{1}{2B} \approx 20.7°$.

Fig. 7.1. The Delaunay refinement algorithm step-by-step with upper bound $B = \sqrt{2}$ on the circumradius-to-shortest-edge ratio. Illustration and most figure texts from Shewchuk [79].

gradients of the interpolated surface. Therefore one should aim at making triangles as equiangular as possible, and this is exactly what is offered by Delaunay triangulations: recall from Chapter 3 that among all possible triangulations of a point set, the Delaunay triangulation is *optimal* in the sense that it maximizes the lexicographical measure of an indicator vector whose elements represent the smallest angle of each triangle. This also implies that the Delaunay triangulation maximizes the smallest interior angle.

The meshing scheme presented in the sequel possesses *mathematical guarantees* for the properties requested above. For example, it offers lower bounds both on the shortest edge and on the smallest interior angle in the mesh (and consequently an upper bound on the largest interior angle), and it offers nice spatial grading from small triangles to larger triangles. Furthermore, the resulting mesh is *size-optimal*: for a given bound on the minimum interior angle, the final mesh has cardinality (number of triangles) within a constant factor of the mesh with the smallest possible cardinality that meets the same angle bound. Size-optimality is not analyzed in this chapter; it is treated thoroughly in [73].

7.3 Node Insertion

We have already proclaimed an incremental meshing approach with "Delaunay refinement" such that new nodes are inserted into the mesh, one at a time, while maintaining a Delaunay triangulation or a constrained Delaunay triangulation (CDT). The starting point is an initial CDT of a PSLG as in the upper leftmost graph in Figure 7.1, with the final mesh after refinement shown in the bottom rightmost figure. Both triangulation of a PSLG and node insertion into a CDT has been considered in depth in Chapter 6. There are some fundamental questions that must be answered in view of the requirements stated in the previous section:

- what exactly is a well-shaped triangle or a triangle with poor quality?
- where should nodes be inserted during the refinement process?
- how can we control termination of the refinement procedure?
- can we control good spatial grading?

Delaunay triangulation of a set of points in the plane was based on one single quantitative measure of the shape of a triangle: the minimum interior angle which was an element in the indicator vector whose lexicographical measure was maximized by the Delaunay triangulation. Delaunay refinement algorithms use the same measure, but for the mathematical analysis we will replace it by an equivalent measure, the *circumradius-to-shortest-edge ratio* r/l, where r is the radius of the circumcircle of a triangle and l the length of the shortest edge of the same triangle. The smallest angle is always opposite

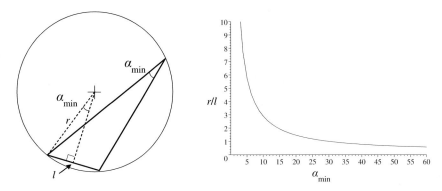

Fig. 7.2. Relationship between circumradius-to-shortest-edge ratio r/l and the minimum angle $\alpha_{\min}$ of a triangle: $r/l = 1/(2\sin\alpha_{\min})$.

to the shortest edge, see Figure 7.2. Basic trigonometry gives this relation between r/l and the minimum angle $\alpha_{\min}$ in a triangle,

$$r/l = \frac{1}{2\sin\alpha_{\min}}. \tag{7.1}$$

Thus, maximizing $\alpha_{\min}$ corresponds to minimizing r/l. A maximum for $\alpha_{\min}$ is possessed by the "perfectly shaped" equilateral triangle where all angles are $60°$, and the corresponding minimum r/l ratio is $1/(2\sin 60°) = 1/\sqrt{3} \approx 0.58$.

Another measure used by many authors is the *aspect ratio* of a triangle. It is defined as the length of the longest edge divided by the length of the shortest altitude of the triangle. For a given minimum angle $\alpha_{\min}$ of a the triangle, the following holds for the aspect ratio A (Exercise 2),

$$\frac{1}{\sin\alpha_{\min}} \le A \le \frac{2}{\tan\alpha_{\min}}.$$

Delaunay refinement algorithms operate with an upper bound B on the circumradius-to-shortest-edge ratio r/l. If all triangles in the final triangulation satisfy $r/l \le B$, then there is no angle α smaller than $\arcsin\frac{1}{2B}$ in the triangulation according to (7.1), and consequently no angle α is larger than $180° - 2\arcsin\frac{1}{2B}$,

$$\arcsin\frac{1}{2B} \le \alpha \le 180° - 2\arcsin\frac{1}{2B}. \tag{7.2}$$

The upper bound B may of course be impossible to obtain if the input PSLG contains angles smaller than $\arcsin\frac{1}{2B}$ between incident constrained edges that must be respected by the final triangulation. We will return to this later.

Any triangle with r/l ratio greater than a required upper bound B will be called *skinny*. The aforementioned works by Ruppert [73] and Chew [15] operate with $B = \sqrt{2}$ and $B = 1$ respectively, which correspond to the following angle bounds.

$B(= \max r/l)$	$\alpha_{\min} = \arcsin \frac{1}{2B}$	$\alpha_{\max} = 180° - 2\alpha_{\min}$
1	30°	120°
$\sqrt{2}$	$\approx 20.7°$	$\approx 138.6°$

With the upper bound $B = 1$, triangles with interior angles smaller than 30° or greater than 120° are interpreted as skinny, while $B = \sqrt{2}$ allows triangles to be more elongated before they are characterized as skinny. The refinement scheme presented later uses the latter bound.

Suppose that the existing triangulation has one or more skinny triangles, for example the triangle t in Figure 7.3(a) with circumradius-to-shortest-edge ratio $r/l \approx 2.14$. Assume for now that there are no constrained triangle edges in the existing triangulation. (The restriction will be removed in the next section.) Recall that a triangle is Delaunay if and only if its circumcircle is empty (Theorem 4.1). As a consequence, it is clear that if a new node is inserted inside the circumcircle of the skinny triangle t, the node will effectively kill t (and any other triangle with circumcircle enclosing the position of the new node).

The new node should not be inserted too close to any existing nodes as this may generate new triangles with short edges and possibly bad shape with dissatisfactory r/l ratio. Intuitively, in order to get new well-shaped triangles by the node insertion, one may argue that the new node should be positioned as far from existing nodes as possible. The position inside the circumcircle of t "as far from existing nodes as possible" is exactly at the center of the circumcircle of t equidistant from the three nodes of t. Hereafter, we call this the *circumcenter* of t. With the definition of Voronoi diagram in mind, and Delaunay triangulation as its dual construction, (Section 3.3 and 3.4), we recall that the circumcenter of t is at the Voronoi point where the Voronoi edges corresponding to t intersect, see Figure 7.3(c). Thus, the circumcenter of t is also more distant from other nodes in the triangulation than it is from the three nodes of t. Figure 7.3(d) shows the Delaunay triangulation after a new node v has been inserted at the circumcenter of t. To simplify the presentation, we assume until further notice that circumcenters of skinny triangles never fall outside the existing triangulation, and thus, new nodes are always inserted outside holes and inside the exterior boundary of the triangulation.

Let us look closer at node insertion at the circumcenter of t and examine its impact on the shape of new triangles. Node insertion into a conventional Delaunay triangulation was treated in detail in Section 4.5 and 4.6, and extended to CTDs in Section 6.6. Figure 7.3(b) shows the influence region R^v of the new node v, that is, the triangles that must be removed since their

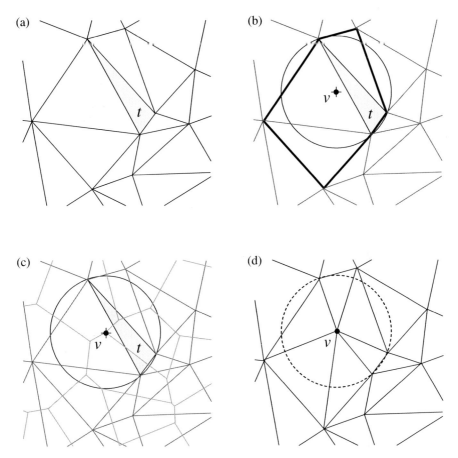

Fig. 7.3. (a): A skinny triangle t in a Delaunay triangulation. (b): t's circumcircle. (c): The Voronoi diagram and the circumcenter of t positioned at a Voronoi point. (d): Updated Delaunay triangulation after insertion of a node at t's circumcenter.

circumcircles enclose v. The new triangulation is obtained by connecting v to all of the boundary nodes of R^v with new triangle edges. (Figure 7.3(d)). Three of the new edges have length equal to the circumradius r of t and the other new edges are longer than r since the circumcircle of the Delaunay triangle t was *node-free* before v was inserted. We say that a circle is node-free if it does not contain any nodes from the triangulation in its interior. (If there is a neutral case, more than three new edges will have length equal to r). Since t was chosen because its r/l ratio was greater than the upper bound B, every new edge has length at least Bl. Thus, if $B > 1$, which corresponds to $\alpha_{min} < 30°$, no new edge ever introduced while "killing" a triangle is shorter than the shortest edge of that triangle. We will return to this fact later when

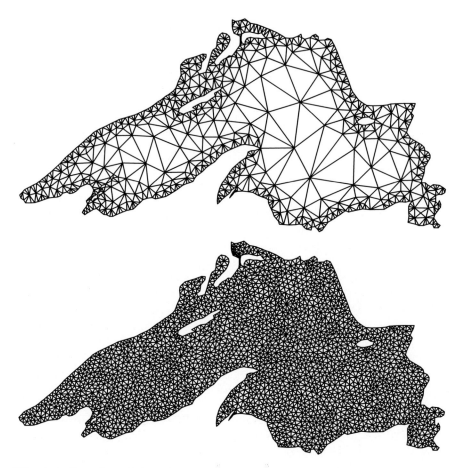

Fig. 7.4. Spatial graded mesh uppermost, and a uniform mesh below. Illustration from Shewchuk [79].

studying the criteria for termination of a Delaunay refinement algorithm. The upper part of Figure 7.4 shows a triangulation where all triangles satisfy an upper bound of $B = 1$. Apart from well-shaped triangles with angle bounds between 30° and 120°, we observe that there is a nice spatial grading from small to large triangles.

Suppose that the definition of a skinny triangle were changed to "*circumradius greater than the shortest edge in the entire mesh*" (used by Chew in [14]). The position of the point insertion is still at the circumcenter of a triangle. If all triangles eventually satisfy this criterion, an upper bound $r/l \leq 1$ would be achieved which corresponds to $\alpha_{\min} \geq 30°$. In addition, this angle bound and the fact that no circumradius of a triangle is greater than the shortest edge in the entire mesh, results in a mesh that must be uniform with all triangles approximately the same size. The number of triangles is, in general, greater than the number of triangles obtained when requiring that $r/l \leq 1$ with r and l from the same triangle. The lower part of Figure 7.4 shows an example where the initial mesh was the same as used for the spatial graded mesh above.

7.4 Splitting Encroached Segments

In the analysis above, we did not account for constrained edges. Constrained edges would not be swapped away during node insertion and could lead to long, thin triangles with small interior angles, and thus dissatisfactory circumradius-to-shortest-edge ratios r/l. We also assumed that the circumcenter of a skinny triangle never fell outside the existing triangulation such that a new node could always be inserted at its circumcenter (thus killing it). In this section we prepare the ground for these assumptions.

To distinguish between constrained edges and other edges of a CDT, we call constrained edges *segments*. The edges constituting the exterior boundary of a CDT are also segments whether they are Delaunay or not, and so are all edges enclosing interior boundaries, or holes. We also say that the triangulation (or the input PSLG) is *segment-bounded* meaning that an exterior boundary, and optionally interior boundaries enclosing holes, are specified and that the edges of the boundaries are segments. Delaunay refinement schemes split segments into shorter contiguous *subsegments* prior to every insertion of a new node as explained next.

Let s be a segment or a *subsegment* of a CDT. The *diametral circle* of s is the circle with center at the midpoint of s that interpolates the endpoints of s. Clearly, the diametral circle is the smallest circle that encloses s. We say that s is *encroached upon* if a node different from the endpoints of s lies on or inside the diametral circle of s and the node is *visible* from the interior of s. The notion of visibility is adopted from constrained triangulations in Section 6.1. Figure 7.5(a) shows a segment that is encroached upon by two nodes.

The first step in the Delaunay refinement process is to eliminate all encroached segments in the CDT. This can be done by recursively splitting them until they are no longer encroached upon. When splitting a segment s, a new node is inserted in its interior and s is replaced by two new segments s_1 and s_2. The new triangulation is updated to be a CDT as described previously in Section 6.6. (See Exercise 7.) Figure 7.5(b) and (c) illustrates this process with the insertion point at the subsegment's midpoint each time. In this example, no subsegment is encroached upon after the second bisection and the recursion stops.

A node that splits an encroached segment s may certainly again encroach upon another segment, say s', and cause s' to be split, and so on. The question

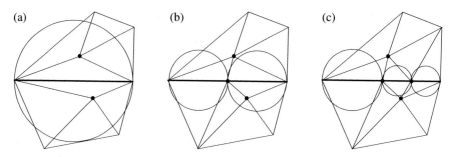

Fig. 7.5. Recursive bisection of a segment that is encroached upon. (a): Two nodes encroach upon the segment initially. (b): After splitting the segment at its midpoint, it is still encroached upon by a node. (c): After the second bisection there is no encroachment.

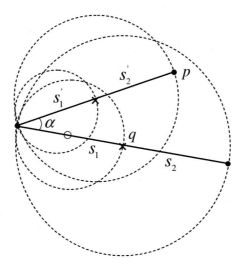

Fig. 7.6. Recursive bisection of incident segments that never terminates when $\alpha \leq 45°$.

is if this process is guaranteed to terminate or if it goes on forever creating shorter and shorter subsegments and smaller triangles.

If s and s' are disjoint without a common endpoint, termination is guaranteed, since the diametral circles of the new subsegments of s sooner or later become small enough so that they do not enclose any nodes of subsegments of s', and vice versa. On the other hand, if s and s' are incident, the situation is different. We say that two segments are *incident* if they share a common node. Assume from now on that encroached segments are always split by recursively bisecting them. Figure 7.6 shows an example where s_1 and s_2 have replaced a segment encroached upon by the node p, and the split node q has again encroached upon a segment that is replaced by s_1' and s_2'. We see that s_1 is again encroached upon and the recursion never terminates. It is easy to verify that if the angle between the two original segments were greater than $45°$, the recursion would stop immediately after q was inserted. Until further notice, let us accept this restriction and assume that no angle between two incident segments in the input PSLG is smaller than or equal to $45°$. (We must also state as a conjecture that there exists no other configuration of segments in the PSLG that may cause a recursion of splits that never terminates.)

Before a node is inserted at the circumcenter of a skinny triangle, there is a "look-ahead" such that segments that *would* be encroached upon by the insertion of the new node are split *prior* to the node insertion. The rationale behind this will be made clear in Section 7.6 when studying conditions for termination of the refinement algorithm presented in the next section. Splitting one segment may cause other segments to be encroached upon and split, but if the angle between incident segments is always greater than $45°$, the segment splitting is guaranteed to terminate. Thus, after splitting segments of the initial CDT, no segment of the triangulation ever gets encroached upon by a new node. Each constraint of the initial PSLG is maintained either as the original constrained edge or as a sequence of contiguous segments during the refinement process.

In the previous section, we assumed that circumcenters of skinny triangles never fell outside the existing triangulation, and thus, new nodes would always be inserted inside the triangulation. We also assumed that a triangle would always be killed by the insertion of a node at its circumcenter. The latter assumption requires that the triangle and its circumcenter are not separated by a segment, that is, a constraint that cannot be swapped away. If they were separated by a segment, edge-swapping would only occur on the same side of the segment as the inserted node, and the triangle would survive. The strategy of splitting all encroached segments prior to any node insertion ensures that these assumptions are legitimate.

Lemma 7.1. *A triangle in a segment-bounded Delaunay triangulation is never separated from its circumcenter by a non-encroached segment.*

Proof. We prove the lemma by contradiction. Suppose that the circumcenter v of a triangle t is separated from t by a segment s and that s is not encroached upon; consult Figure 7.7.

Let p be any point in the interior of t. Let s be the segment closest to p that intersects $\overline{pv}$, and s' the part of s inside the circumcircle of t. Let $C(t)$, $C(s)$ and $C(s')$ denote the circumcircle of t, and the diametral circles of s and s' respectively. Because t is (constrained) Delaunay, both endpoints of s are outside $C(t)$. Since v and t lie on opposite sides of s', the portion of $C(t)$ on the same side of s' as t lies inside $C(s')$.

Then at least one node of t must be strictly inside $C(s')$. (One or two nodes of t may coincide with endpoints of s'.) But $C(s')$ is inside $C(s)$, so

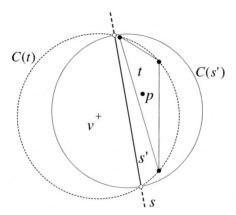

Fig. 7.7. Illustration for Lemma 7.1.

then at least one of the nodes of t encroaches upon s and we have reached a contradiction. □

This result ensures that the visibility between the circumcenter of a triangle and the triangle itself is never obscured by any segment after all encroached segments have been split. Therefore, a triangle will always disappear when a node is inserted at its circumcenter. Since edges of the exterior boundary and edges of interior boundaries enclosing holes are also segments, a triangle's circumcenter never falls outside the triangulation or inside holes where there are no triangles. So, when there are no encroached segments present in the triangulation, a new node can always be inserted at the circumcenter of a skinny triangle, and the skinny triangle will be killed by this operation. This is the crux of the refinement algorithm that follows below.

It is also worthwhile to notice here that if a CDT has no encroached segments and the only boundary of the triangulation is the convex hull of its nodes, then the triangulation is in fact a conventional Delaunay triangulation and not only a CDT. (See Exercise 4.)

7.5 The Delaunay Refinement Algorithm

Let us put the different pieces together in a *Delaunay refinement algorithm*. Let the geometric input to the algorithm be a segment-bounded PSLG. Normally one must also provide some additional information about which sequences of segments that define boundaries. If the triangulation domain is connected, there is exactly one exterior boundary and optionally interior boundaries enclosing holes that should not be covered by triangles in the final mesh. Figure 7.1 shows a companion example for the Algorithm 7.1.

Apart from constrained triangulation in Step 1 that creates an initial CDT, there are two basic operations in the algorithm:

- `SplitSegment(s)`:
  Insert a node *at the midpoint* of a segment s as explained in the previous section.
- `KillTriangle(t)`:
  Insert a node at the circumcenter of a (skinny) triangle t as explained in Section 7.3.

`KillTriangle` is only applied when there are no encroached segments present. Both functions update the triangulation to be constrained Delaunay. The circle criterion ensures that t is removed by `KillTriangle` and never occurs again in the triangulation.

Algorithm 7.1 Delaunay refinement

```
 1. Make the initial CDT of the PSLG.
    Remove triangles outside the triangulation domain.
 2. while skinny triangles remain      // (controls termination)
 3.    while any segment s is encroached upon
 4.       SplitSegment(s)
 5.    Let t be a skinny triangle and v the circumcenter of t.
 6.    if v encroaches upon any segments s₁, s₂, ..., sₖ      // "look-ahead"
 7.       for i = 1, ..., k
 8.          SplitSegment(sᵢ)
 9.       goto 3
10.    else
11.       KillTriangle(t)
12.       goto 5
```

$ $5.$ Let t be a skinny triangle and v the circumcenter of t.
6. **if** v encroaches upon any segments $s_1, s_2, \ldots, s_k$
7. **for** $i = 1, \ldots, k$
8. `SplitSegment`(s_i)
10. **else**
11. `KillTriangle`(t)

Construction of the initial CDT in Step 1 can be accomplished by the algorithms presented in Chapter 6. The incremental approach, which entails first making a conventional Delaunay triangulation from all nodes in the PSLG and then inserting all constraints, results in a CDT with convex boundary and no holes. Therefore triangles outside the exterior boundary and inside interior boundaries defining holes must be removed. The outer loop from Step 2 starts by splitting all encroached segments and then picks an arbitrary skinny triangle t. If the circumcenter of t encroaches upon any segments, then segment splitting continues until no segment is encroached upon.

The skinny triangle t may disappear already during segment splitting in Step 8 if one of its edges is swapped. Also, new nodes that are used for segment splitting may encroach upon other segments. Therefore Step 9 takes us back to segment splitting again in Step 3.

After a new node has been inserted at the circumcenter of a triangle in Step 11, there are still no segments encroached upon. This is ensured by the "look-ahead" in Step 6 that rejects insertion of a new node if it would encroach upon a segment. Therefore, KillTriangle is proceeded by picking a skinny triangle again in Step 5.

In general, the final triangulation produced by Algorithm 7.1 depends on the specific order in which encroach subsegments and skinny triangles are chosen. If one wants to minimize the number of triangles in the final mesh, one may consider a specific ordering of these operations. For example, one may maintain a priority queue of skinny triangles ordered by their r/l ratio and always kill the skinniest triangle first. Although no guarantee can be given, one may expect to get fewer triangles in the final mesh with this strategy especially if the upper bound B on the circumradius-to-shortest-edge ratio is set low [77].

The stop criterion for Algorithm 7.1 is only linked to the presence of skinny triangles as defined by the upper bound on the circumradius-to-shortest-edge ratio. Even though skinny triangles have been removed upon termination of the algorithm (if it terminates), there may be triangles left that are too large to satisfy small error bounds for a FEM problem or an interpolation problem. The mesh may be refined further, for example to meet upper bounds on the area of triangles that may vary over the domain.

One may of course alter the meaning of skinny triangle, and thus, which triangles are removed by KillTriangle. For example, in Section 7.3 we argued that if a skinny triangle is one with circumradius greater than the shortest edge in the entire mesh, the final mesh would be uniform with all triangles approximately the same size as the mesh in the lower part of Figure 7.4.

Although each skinny triangle t that is chosen in Step 5 is guaranteed to be killed by KillTriangle or SplitSegment, these operations may also create new skinny triangles. This brings us to the question of termination of the algorithm. Will the algorithm run forever and never get rid of skinny triangles? Or, is termination guaranteed under certain conditions, for example, if the upper bound B of the circumradius-to-shortest-edge ratio, which defines a skinny triangle, is not set too low? The first scenario would definitely result in an endless loop with smaller and smaller edges and triangles inside the triangulation domain. Another question related to termination is how one should deal with the restriction of the previous section where we assumed that no angles between two adjacent segments in the input PSLG are smaller than 45°. Without this assumption, the loop in Step 3 and 4 would be endless if this occurred. We will address these questions in the following sections.

7.6 Minimum Edge Length and Termination

The refinement part of Algorithm 7.1 has two basic operations, KillTriangle
and SplitSegment that both insert new nodes into the triangulation. Control
statements of the algorithm control the interplay between them such that
each KillTriangle is done after a sequence of SplitSegment operations has
removed all encroached subsegments. The question of termination is if and
how the algorithm can reach a state where there are no skinny triangles left
in the triangulation. This section analyzes the refinement process in view of
these questions.

It is intuitively clear that the distribution and the density of nodes and
segments of the input PSLG has an impact on the size of triangles and edges,
and thus on the number of triangles produced by Algorithm 7.1. To formalize this, we first define a function in the plane that measures the density of
incoming nodes and edges.

Definition 7.1 (Local feature size). *Given a PSLG G. The* local feature
size *at an arbitrary point* $p \in R^2$ *relative to G, denoted* $lfs_G(p)$, *or only* $lfs(p)$,
*is the radius of the smallest disk centered at p that intersects two non-incident
nodes or segments of G.*

A disk intersects a point if the point lies inside the circumscribed circle
of the disk. Figure 7.8 illustrates the local feature size for some points in the
plane relative to a given PSLG. Note that $lfs(p)$ is never zero for any point
p since the disk centered at p, which determines $lfs(p)$, must intersect *non-
incident* features. It is also emphasized that $lfs(\cdot)$ is defined relative to the

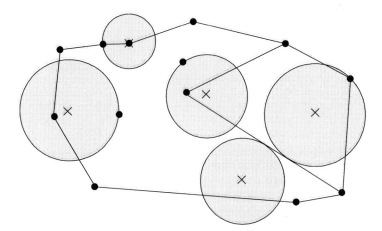

Fig. 7.8. Illustration of local feature size $lfs()$ at some points relative to a planar
straight line graph. An arbitrary point in the plane marked with × has local feature
size equal to the radius of the circle drawn with center at ×.

input PSLG only; it is not changed when making the initial CDT or when nodes are inserted during the refinement process. It is easy to see that $lfs(\cdot)$ is continuous everywhere in the plane. It is left to the reader to prove that its steepness is bounded in the range $[-1, 1]$ in any direction as expressed by this lemma[1].

Lemma 7.2. *Given a PSLG and two points p and q in the plane. Then*

$$lfs(q) \leq lfs(p) + d(p, q),$$

where $d(p, q)$ is the Euclidean distance between p and q.

In addition to local feature size, which gives the density of nodes and segments of the input PSLG, we also need a measure of density of nodes as they are inserted during the refinement process. The length of the shortest edge connected to a new node is a natural measure of the density at that node.

Definition 7.2 (Insertion radius [80]). *Let v be a node inserted at the circumcircle of a skinny triangle or at the midpoint of an encroached subsegment. The* insertion radius r_v *of v is the length of the shortest Delaunay edge connected to v immediately after* KillTriangle *or* SplitSegment *has inserted v and updated the triangulation to be constrained Delaunay.*

The definition is also applied to nodes that are not actually inserted during the refinement process:

- An *input node* v has insertion radius equal to the length of the shortest Delaunay edge connected to v after the initial CDT has been constructed and before the refinement process starts.
- A *rejected node* v, which encroaches upon a segment (Step 6 of Algorithm 7.1), has insertion radius equal to the shortest Delaunay edge connected to v *as if* v had been inserted into the CDT.

Recall that all new edges created when inserting a node v into a CDT have v as a common endpoint, see Figure 7.3(d). The Delaunay property ensures that the closest node visible from v defines the shortest edge connected to v.

We will now look at sequences of node insertions at circumcenters of skinny triangles and at midpoints of encroached subsegments and see if, and under what conditions, Algorithm 7.1 converges such that all skinny triangles eventually are removed. When studying the insertion of a new node v by KillTriangle or SplitSegment, in each of several (all possible) cases we will look for a lower bound on the insertion radius r_v either in terms of the local feature size at v, or as a constant factor multiplied by the insertion radius r_p

---

[1] The result is not used in further analyses in this chapter.

of a previous inserted node p near v. That is, we want to find out if the shortest edge created when a node is inserted is longer than a bound determined by the input PSLG or longer than an already existing edge in the mesh. The node p is called the *parent* of v and can be regarded as the node that "is responsible for the insertion of v". The idea is as follows:

If such a bound on the shortest edge connected to the insertion node v is found, then we can avoid creating a chain of edges with ever-decreasing lengths that would cause Algorithm 7.1 to run forever in an endless loop without getting rid of skinny triangles.

Below are the possible cases that can occur under the functions `KillTriangle` and `SplitSegment` and all the different roles of the parent node p.

Case 1. (`KillTriangle`) *v is inserted at the circumcenter of a skinny triangle t.*
The parent node p is chosen to be one of the two endpoints of the shortest edge of t. (Figure 7.9).

Case 2. (`SplitSegment`) *v is a node inserted at the midpoint of a segment s that is encroached upon by a node p.*
Thus p lies inside the diametral circle of s. If more than one node encroaches upon s, assume without loss of generality that p is the closest node to v that encroaches upon s. The shortest edge connected to v, which defines the insertion radius r_v, has p as the other endpoint unless p is not yet inserted and

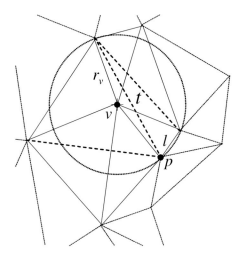

Fig. 7.9. Case 1 where v is inserted at the center of a skinny triangle t. The parent node p is chosen as one of the endpoints of the shortest edge of t.

thus rejected. This follows from the Delaunay property. Four possible roles of p must be considered under `SplitSegment`.

> **Case 2a.** *p is an input node, or p is a node inserted in a segment not incident to s. (Figure 7.10(a))*

> **Case 2b.** *p is at the circumcenter of a skinny triangle and thus rejected since it encroaches upon s. (Figure 7.10(b), and Step 6–8 of Algorithm 7.1.)*

> **Case 2c.** *p is a node on a segment s' incident to s that makes an angle $45° \leq \alpha < 90°$ with s. (Figure 7.10(c))*

> **Case 2d.** *As Case 2c with $\alpha \leq 45°$. (Figure 7.10(d))*

Concerning the last two cases, if α were greater than $90°$, p would not encroach upon s; and $\alpha \leq 45°$ was excluded already in Section 7.4 due to recursive bisection of s and s' that never terminates. We formalize lower bounds on r_v in the different cases in the following lemma.

Lemma 7.3. *The insertion radius r_v has the following lower bounds in the different cases above:*

Case 1. $r_v \geq Br_p$,
Case 2a. $r_v \geq lfs(v)$,
Case 2b. $r_v \geq r_p/\sqrt{2}$,
Case 2c. $r_v \geq r_p/(2\cos\alpha)$,
Case 2d. $r_v \geq r_p \sin\alpha$.

Proof.

Case 1. Consult Figure 7.9. This case was analyzed in Section 7.3 where we found that $r_v \geq Bl$, where l is the length of the shortest edge of the skinny triangle t. But $l \geq r_p$, where r_p is the insertion radius of p, so $r_v \geq Br_p$.

Case 2a. Consult Figure 7.10(a). Since both p and v fall on the input PSLG, an upper bound on r_v can be expressed in terms of the local feature size at v. It follows directly from the definition of local feature size that $r_v \geq lfs(v)$.

Case 2b. Consult Figure 7.10(b) and (e). Since p is rejected, the insertion radius r_v is $|s|/2$, which is the length of a subsegment created when v is inserted at the midpoint of s. The circle criterion ensures that the circumcircle of t cannot contain an endpoint of s in its interior. The insertion radius r_p of p is the circumradius of t. A lower bound on r_v in terms of r_p is obtained when r_v/r_p is minimized, and thus when r_p is maximum under the constraint

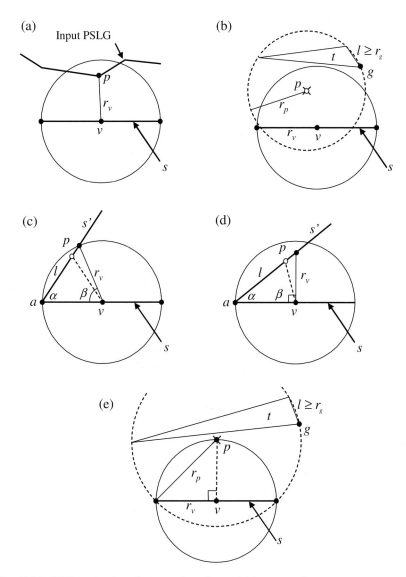

Fig. 7.10. Different roles of a parent node p which encroaches upon a segment s.
(a): p is on the input PSLG.
(b): p is at the circumcenter of a skinny triangle (and rejected).
(c) and (d): p is on a segment s' incident with s with $\alpha \geq 45°$ and $\alpha < 45°$ respectively. The filled bullet • at p indicates the position of p which defines the lower bounds on the insertion radius r_v in each case of Lemma 7.3.
(e): illustrates the same case as (b), but with p in a position which defines the lower bound on r_v.

that p lies on or inside the diametral circle of s. Figure 7.10(e) illustrates the position of p and the circumcircle of t when this occurs. This gives the lower bound $r_v \geq r_p/\sqrt{2}$.

Case 2c. Consult Figure 7.10(c). Let a be the apex of the acute angle between s and s' and l the length of segment $\overline{ap}$; then $l \geq r_p$. Because p is on or inside the diametral circle of s, r_v is equal to the length of segment $\overline{vp}$. The law of sines gives $r_v/l = \sin\alpha/\sin\beta$, where β is the angle $\angle avp$ at v. Since $l \geq r_p$, we can find a lower bound on r_v for a fixed α by minimizing the fraction r_v/l under the constraint that p lies on or inside the diametral circle of s. Because $\alpha \geq 45°$, the angle β is at most $90°$ and thus $\sin\beta \leq 1$. So, r_v/l is minimized when β is maximum under the given constraint. This happens when p lies on the diametral circle of s as indicated in the figure, and this gives $r_v = l/(2\cos\alpha)$. Thus, $r_v \geq r_p/(2\cos\alpha)$.

Case 2d. Consult Figure 7.10(d). The same analysis as in Case 2c can be used. But now β is not bounded as in Case 2c, so $r_v/l = \sin\alpha/\sin\beta$ is minimized when $\sin\beta = 1$ and thus $\beta = 90°$. (The point p is still inside the diametral circle of s.) We get $r_v = l\sin\alpha$, and the lower bound $r_v \geq r_p\sin\alpha$.   □

Let us bring each case a step further and determine what values of B and α guarantee a lower bound on the shortest edge created in the refinement process. The following analysis concludes with a theorem that brings us even closer to termination of Algorithm 7.1.

Recall from Section 7.4 that **Case 2d** generates an endless sequence of new edges with decreasing lengths. (Endless loop with `SplitSegment` in Step 3 and 4 of Algorithm 7.1.) Even if $\alpha > 45°$, which gives termination when two incident segments are viewed separately, the shortest edge connected to v may be shorter than the shortest edge connected to p as expressed by the lower bound $r_v \geq r_p/(2\cos\alpha)$ in **Case 2c**. When this case is intermingled with the other cases, we may also get an endless sequence of decreasing edges. But if $\alpha \geq 60°$, then $r_v \geq r_p$, and in Case 2c the length of the shortest edge connected to v will always be longer than the shortest edge connected to p when p was inserted.

In **Case 2b**, with $r_v \geq r_p/\sqrt{2}$, the lower bound on r_v is *not in terms of a node that is actually inserted into the mesh.* And here lies the rationale for the "look-ahead" mechanism in Step 6–8 of the algorithm. The node p, which is rejected, is at the circumcenter of a skinny triangle and itself has a lower bound on its insertion radius from the analysis of `KillTriangle` in Case 1. Let g denote the parent of p (g for "grandparent" of v), see Figure 7.10(b). Then, from Case 1 we have $r_p \geq Br_g$, and combined with Case 2b we get $r_v \geq r_p/\sqrt{2} \geq Br_g/\sqrt{2}$. Hence, when $B \geq \sqrt{2}$ we obtain $r_v \geq r_g$ which guarantees that the shortest edge delivered by Case 2b is longer than the shortest edge from the insertion of g. This bound on B is of course also sufficient for `KillTriangle` in **Case 1** which now gives $r_v \geq \sqrt{2}r_p$.

Case 2a with $r_v \geq lfs(v)$ bounds the length of the shortest edge connected to v to the geometry of the input PSLG. So, in this case, the shortest edge connected to v cannot be shorter than $\min lfs(\cdot)$ over the entire domain. Since v and p lie on non-incident segments (or p is an isolated point on the input PSLG), this bound can be improved to

$$r_v \geq d_{\min}, \tag{7.3}$$

where $d_{\min}$ is the shortest distance between two non-incident entities (nodes or segments) of the input PSLG[2]. Clearly $d_{\min} \geq \min lfs(\cdot)$, so the latter is indeed a stronger bound on r_v. The same bound is of course valid for edges of the input PSLG; an input edge can never be shorter than $d_{\min}$ since its endpoints are non-incident nodes of the input PSLG.

Case 2a is the only case that bounds r_v to the geometry of the input PSLG. Provided that $B \geq \sqrt{2}$ and the smallest angle between incident segments of the input PSLG is at least $60°$, the other cases have bounds at least $r_v \geq r_p$ or $r_v \geq r_g$, where p or g is a node that already exists when v is inserted. Thus, in each of these cases, the length of the shortest edge created is at least the length of an existing edge in the mesh. Case 2a therefore represents a "last barricade" and bounds the length of the shortest edge that can ever be created by Algorithm 7.1 to $d_{\min}$. We prove this formally by "tracking the history back in time" from an insertion node v to nodes that have been inserted prior to v.

Theorem 7.1. *Let the angle between two incident segments of the input PSLG be at least $60°$, and let the upper bound on the circumradius-to-shortest-edge ratio B be at least $\sqrt{2}$. Then the shortest edge created by Algorithm 7.1 has length at least $d_{\min}$, where $d_{\min}$ is the shortest distance between two non-incident entities (nodes or segments) of the input PSLG.*

Proof. We prove the theorem by contradiction. Clearly, any edge between nodes of the input PSLG must have length at least $d_{\min}$. Suppose that an edge e with length $r_v < d_{\min}$ is created while inserting a node v. Without loss of generality, assume that e is the first edge created that is shorter than $d_{\min}$. Since $r_v < d_{\min}$, v cannot belong to Case 2a (cf. (7.3)), and case 2d is omitted due to the restriction on input angles that are at least $60°$. In each of the other cases, $r_v \geq r_p$, where p is a node that exists in the mesh when v is inserted. By the assumption that e is the first edge introduced with length less than $d_{\min}$, we must have $r_p \geq d_{\min}$. But then $r_v \geq d_{\min}$, which is a contradiction. $\square$

To summarize the analysis so far: if the acute angle α between incident segments of the input PSLG is at least $60°$ and the upper bound B on the

---

[2] Shewchuk uses $lfs_{\min}$ in [80], though this measure is not a lower bound on $lfs(\cdot)$, see Exercise 6 and 7.

circumradius-to-shortest-edge ratio is at least $\sqrt{2}$, then Algorithm 7.1 cannot produce edges with ever-decreasing lengths linked to the new insertion nodes; and there is a lower bound $d_{\min}$ on the shortest edge in the final mesh that depends on the geometry of the input PSLG.

But is this sufficient for termination of the algorithm? The Delaunay property helps us further: since every edge created by KillTriangle and SplitSegment is Delaunay in a CDT, the shortest edge connected to an insertion node v has the closest node visible from v as its other endpoint. Then a new node v can never be inserted closer to an existing node p than the lower bound $d_{\min}$ on the shortest edge, unless the visibility between v and p is obscured by a segment.

Imagine that a circle with radius $d_{\min}$ is drawn with center at each new node v. Each such circle is node-free initially and a node inserted later than v cannot fall inside the circle around v if that node is visible from v. Even though these circles may overlap, they cannot coincide. Sooner or later there will be no space left to insert new nodes and draw such node-free circles, so the process must terminate.

When the algorithm terminates, no interior angle is smaller than $\arcsin \frac{1}{2B} \approx 20.7°$ and no angle is greater than $180° - 2 \arcsin \frac{1}{2B} \approx 138.6°$ (inequality 7.2). In many applications this may suffice, but there may certainly also be a need for removing triangles that are less skinny. For practical applications, one should consult the aforementioned works by Chew [15] and Shewchuk [79, 80] for variations to Algorithm 7.1 that guarantee larger interior angles. The next section looks at the $60°$ angle restriction of the input PSLG under Case 2c.

7.7 Corner-Lopping for Handling Small Input Angles

A serious limitation of Algorithm 7.1 is the restriction that acute angles between incident segments of the input PSLG must be at least $60°$. Any meshing algorithm used in practice, for example for creating FEM meshes, must accept smaller angles. Even arbitrary small input angles must be accepted in many applications. This section gives a practical workaround due to Bern, Eppstein and Gilbert [9] and Ruppert [73], called *corner-lopping* that deals with arbitrary small input angles. There are other methods that give more well-shaped meshes and are easier to implement, but we demonstrate corner-lopping here because it is conceptually very simple. The method presented in [63] is easier to implement and delivers meshes with good spatial grading in areas with small input angles.

Figure 7.11, upper left, shows a PSLG with four angles smaller than $60°$ between incident segments. There is no guarantee for termination of Algorithm 7.1 with this PSLG as input. The simple idea of corner-lopping is to "lop off" every such sharp corner prior to Delaunay refinement and replace the

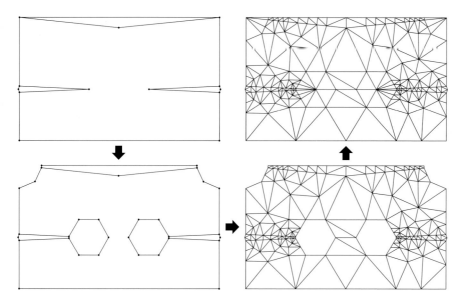

Fig. 7.11. Corner-lopping when angles of the input PSLG are less than 60°. Illustration from Shewchuk [80].

input PSLG G with a modified PSLG G' without any angle smaller than 60°. Here is a brief procedural description of Delaunay refinement with corner-lopping. Let a denote the apex of an angle smaller than 60° in G.

1. Draw a circle C (*lopping circle*) with center at a and radius r. Choose r such that there is no overlapping with other lopping circles. If r is a fraction of the local feature size at a, say $r = lfs_G(a)/3$, then C is separated from all other lopping circles by a distance at least $lfs_G(a)/3$ (Figure 7.11, lower left).
2. Insert new nodes where C intersects segments of G; thereby dividing C into two or more arcs. Angles at a (spanned by circle arcs that fall inside the triangulation domain) are split by inserting nodes at the circle arcs until no angle is smaller than 60°
3. Insert new segments between new nodes on C inside the triangulation domain (*shield segments*). Remove the node at a and subsegments incident with a leaving a modified PSLG G'. The area of the lopped corner now falls outside the new triangulation domain; it is either a hole with an interior closed boundary, or it is outside the new exterior boundary.
4. Run Algorithm 7.1 on G' with an upper bound on the circumradius-to-shortest-edge ratio $B \geq \sqrt{2}$. Since there are no angles smaller than 60° between incident segments, we have a mathematical guarantee for

termination of Algorithm 7.1 with G' as input. The refined mesh has minimum angle $\alpha_{\min} \geq \arcsin \frac{1}{2B}$ ($\approx 20.7°$). (Figure 7.11, lower right).

5. Insert the node at apex a again and a fan of segments (*spoke edges*) between a and nodes on the shield edges. (Figure 7.11, upper right).

Note that, due to encroachment, shield segments may have been subdivided further by the refinement algorithm. Hence, there may be several small angles, and thus skinny triangles, at a defined by spoke edges and shield segments. One may process the lopped areas further to improve the mesh with fewer small angles, but such schemes may be quite involved.

While the basic Delaunay refinement algorithm is easy to implement (provided that an algorithm for constructing the initial CDT is available) corner-lopping requires more effort. In the lopped area, the quality of the mesh heavily relies on the radius of the lopping circle and on how (and if) the lopped area with spoke edges is modified as a last step to get fewer small angles.

7.8 Spatial Grading

Intuitively, the refinement scheme presented in the preceding sections gives good spatial grading from small to large triangles in the mesh. This can be understood by the lower bounds on the length of the shortest edge created when a new node v is inserted. As Lemma 7.3 shows, the bounds are in terms of the local feature size at v, or in terms of the length of an existing edge near v. If $B \geq \sqrt{2}$ and $\alpha \geq 60°$, the shortest edge created at v is greater than a nearby existing edge, apart from the encroachment Case 2a which determines the shortest edge in terms of the local feature size at v. So, apart from Case 2a, a short edge cannot be created at v if there is no edge in the neighborhood of v that is even shorter. And, in Case 2a a short edge can only be created at v if the distance between non-incident entities of the input PSLG is small near v (cf. (7.3)). A formal mathematical analysis of spatial grading can be found in [73] and [80].

7.9 Exercises

1. There are many interactive tools for Delaunay triangulation. Use one of these, for example *VoroGlide*, to refine a mesh by inserting new nodes at circumcircles of skinny triangles. (Search for it on the Internet.) Do this several times to get a "feeling" of how the refinement process behaves. Try also to find an interactive tool that handles constraints and run Algorithm 7.1 interactively. (Alternatively and even better, implement your own interactive meshing program using TTL.)

2. Let A be the aspect ratio of a triangle, i.e., the length of the longest edge divided by the shortest altitude, and α_{min} the minimum angle of the triangle. Show that the following holds:

$$\frac{1}{\sin \alpha_{min}} \leq A \leq \frac{2}{\tan \alpha_{min}}.$$

3. Verify that splitting of incident segments termintes if the angle between the segments is greater than 45°.

4. Show that if no segments (or subsegments) in a CDT are encroached upon, then the CDT is a conventional Delaunay triangulation:

   a) Show that the non-encroached segments (or subsegments) are true Delaunay edges.

   b) Show that the other edges are true Delaunay edges.

5. Prove Lemma 7.2.

6. Draw an example which shows that d_{min} in Section 7.6 is not a lower bound on $lfs(\cdot)$ over the entire domain.

7. Show that $d_{min} = \min_u lfs(u)$, where u is a node in the input PSLG.

8. Use TTL to program Algorithm 7.1. (The initial CDT of the input PSLG can be created by built-in functions in TTL). Assume that the smallest angle between two incident segments of the input PSLG is at least 60° such that corner-lopping is not necessary (see also the next exercise).
   Run the algorithm with different upper bounds B on the circumradius-to-shortest-edge ratio and report the results. Try also $B < \sqrt{2}$ which is below the theoretical minimum for which the algorithm is guaranteed to terminate.

9. Make options with variations to the algorithm from the exercise above.

   a) Pick skinny triangles using a priority queue (see Section 7.5).

   b) Define a skinny triangle to be one with circumradius greater than the shortest edge in the entire mesh, and observe how this influences the spatial grading of the final mesh.

10. Extend the software from the exercises above with (naive) corner-lopping.

Least Squares Approximation
of Scattered Data

Surface reconstruction from scattered data in applications like reverse engineering, geographic information systems and geological modeling usually involves huge amount of data. Creating surface triangulations with vertices in every given data point may be too memory demanding and time consuming, and is usually not required in such cases. Data acquired with scanning devices or data from seismic surveys may also contain considerable noise. Therefore a "good" approximation to the measured data is often sought instead of exact interpolation. Approximation by least squares is a common approach to constructing surfaces from measured data and in this chapter least squares approximation is applied to fit surface triangulations to data.

We first cover the theory of least squares approximation over triangulations including practical and theoretical aspects related to smoothing, weighting and constraints. At the end of the chapter, we present an adaptive multilevel approximation scheme for huge surface models based on a dynamic triangulation structure called *binary triangulations*. Several numerical examples are given based on this scheme.

8.1 Another Formulation of Surface Triangulations

In this chapter, we restrict a surface triangulation f to the finite dimensional space $S_1^0(\Delta)$ of piecewise linear polynomials over a triangulation Δ, i.e.,

$$S_1^0(\Delta) = \left\{ f \in C^0(\Omega) : f|_{t_i} \in \Pi_1, \ i = 1, \ldots, |T| \right\}, \tag{8.1}$$

where Π_1 is the space of bivariate linear polynomials, t_i is a triangle in Δ, $|T|$ is the number of triangles in Δ, and Ω is the 2D domain triangulated by Δ. (See also Chapter 1). This is the function space that was introduced in Chapter 1 and used for data dependent triangulations in Chapter 5, but here we will represent f as a *linear combination* of *basis functions* that span the space $S_1^0(\Delta)$. As a basis we use compactly supported piecewise linear pyramidal basis functions

$$(N_1(x,y), N_2(x,y), \ldots, N_n(x,y)),$$

where n is the number of vertices in the triangulation Δ. The basis functions satisfy

$$N_i(v_j) = \delta_{ij}, \quad j = 1, \ldots, n, \tag{8.2}$$

where $v_j = (x_j, y_j)$ is a vertex in the underlying triangulation and δ_{ij} is the Kronecker delta function which is 1 for $i = j$ and 0 otherwise. Figure 8.1 shows an example of a basis function $N_i(x,y)$. The support Ω_i of $N_i(x,y)$ is the union of those triangles that have $v_i = (x_i, y_i)$ as a common vertex. Each basis function is greater than zero strictly inside its support and zero elsewhere. The basis functions are also *linearly independent* on Ω. This is an important (and necessary) property for the approximation schemes presented later in this chapter. Linear independency implies that for $\sum_{i=1}^n \alpha_i N_i(x,y)$ to be zero for all $(x,y) \in \Omega$, then all the real values α_i, $i = 1, \ldots, n$ must be zero.

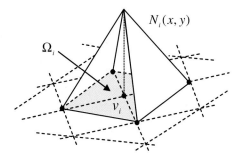

Fig. 8.1. A basis function $N_i(x,y)$ for the function space $S_1^0(\Delta)$ and its (compact) support Ω_i.

A function $f(x,y)$ in $S_1^0(\Delta)$, representing a surface triangulation, can be written as a linear combination of the pyramidal basis functions,

$$f(x,y) = \sum_{i=1}^n c_i N_i(x,y), \tag{8.3}$$

where the real values $c_1, \ldots, c_n$ are coefficients of f. We also call $\mathbf{c} = (c_1, \ldots, c_n)^T$ the *coefficient vector* of f. Only three basis functions can be non-zero strictly inside a triangle. Therefore the sum in (8.3) for any point (x,y) in the plane has at most three non-zero terms. The actual evaluation of (8.3) can of course still be done by barycentric coordinates as explained in Section 1.6. By the property in (8.2), we see that when evaluating f at any triangle vertex $v_j = (x_j, y_j)$ of Δ, we get $f(x_j, y_j) = c_j$. Thus, the coefficient vector $(c_1, \ldots, c_n)$ equals the data vector $(z_1, \ldots, z_n)$ if f is an interpolating surface.

For later use, we derive the gradient of a function $f \in S_1^0(\Delta)$ at a point (x, y) inside a triangle t. Let g be the restriction of f to t. That is, g is a linear triangle patch in 3D space spanned by the vertices of t. The gradient of g is

$$\nabla g = \left(\frac{\partial g}{\partial x}, \frac{\partial g}{\partial y} \right).$$

Since g is a linear polynomial, the gradient is constant inside t. The gradient also determines the normal vector of the triangle which can be written

$$\mathbf{n} = \left(-\frac{\partial g}{\partial x}, -\frac{\partial g}{\partial y}, 1 \right). \tag{8.4}$$

Formally, the partial derivatives $\partial g/\partial x$ and $\partial g/\partial y$ at a point (x, y) are found by taking derivatives term by term (of the three non-zero terms) of (8.3). Here we derive the partial derivatives geometrically from the normal vector given by (8.4).

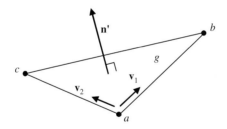

Fig. 8.2. Triangle patch and normal vector.

Let the triangle t have vertex indices (a, b, c) in counterclockwise ordering, see Figure 8.2. The 3D coordinates associated with the vertices are (x_a, y_a, z_a), (x_b, y_b, z_b), and (x_c, y_c, z_c). The normal vector to g is the cross product between the vectors $\mathbf{v}_1 = (x_b - x_a, y_b - y_a, c_b - c_a)$ and $\mathbf{v}_2 = (x_c - x_a, y_c - y_a, c_c - c_a)$ and can be written (Exercise 1),

$$\mathbf{n}' = \mathbf{v}_1 \times \mathbf{v}_2 = (-\eta_a c_a - \eta_b c_b - \eta_c c_c, \ -\mu_a c_a - \mu_b c_b - \mu_c c_c, \ 2A) \tag{8.5}$$

where

$$\eta_a = (y_b - y_c), \quad \eta_b = (y_c - y_a), \quad \eta_c = (y_a - y_b),$$
$$\mu_a = (x_c - x_b), \quad \mu_b = (x_a - x_c), \quad \mu_c = (x_b - x_a),$$

and A is the area of the triangle t (in the xy-plane),

$$A = \frac{1}{2} \left((x_b - x_a)(y_c - y_a) - (y_b - y_a)(x_c - x_a) \right). \tag{8.6}$$

If the normal vector in (8.5) is scaled by a factor $1/(2A)$, we see from (8.4) that the first two vector components become $-\partial g/\partial x$ and $-\partial g/\partial y$. Thus, the gradient of g is

$$\nabla g = (\eta_a c_a + \eta_b c_b + \eta_c c_c,\ \mu_a c_a + \mu_b c_b + \mu_c c_c)/2A. \tag{8.7}$$

Note that the gradient is expressed as a linear combination of the three coefficients corresponding to the vertices of the triangle t. For later use, when constructing smoothing operators for our approximation schemes, we rewrite this slightly such that the partial derivatives are expressed as linear combinations of all the coefficients of f,

$$\nabla g = \frac{1}{2A} \left(\sum_{i=1}^{n} \eta_i c_i,\ \sum_{i=1}^{n} \mu_i c_i \right). \tag{8.8}$$

Here η_i and μ_i are non-zero only when i is a vertex index of the triangle t, that is, a, b and c in our example above.

8.2 Approximation over Triangulations of Subsets of Data

While the surface triangulation f defined above was an interpolating surface in the sense that $f(x_k, y_k) = z_k$ for all points of the given data $\{(x_k, y_k, z_k)\}_{k=1}^{n}$, we now assume that the number of given data points is larger, usually much larger, than the number of vertices in the triangulation Δ, and that we want to find a surface triangulation f defined over Δ that is an *approximation* to all the given data.

Given a set of distinct non-collinear points $P = (x_1, y_1), (x_2, y_2), \ldots, (x_m, y_m)$ in the plane and corresponding data values $z_1, z_2, \ldots, z_m$, let Δ be a triangulation of a subset V which contains $n \leq m$ vertices from P. Assume without loss of generality that the vertices of V are the n first points of P and that the numbering of these vertices is the same as in P,

$$V = \{v_k = (x_k, y_k)\}_{k=1}^{n}. \tag{8.9}$$

We also assume that all points in $P \setminus V = (x_{n+1}, y_{n+1}), \ldots, (x_m, y_m)$ fall (strictly) inside Δ. (See Figure 8.3.) Since $n \leq m$, and since there is strict inequality in most practical cases, with many more points in P than there are degrees of freedom (coefficients of f), it is not possible in general to

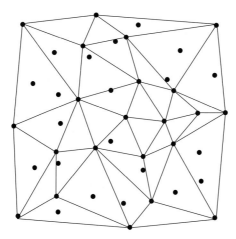

Fig. 8.3. Triangulation generated from a subset of a given data set.

find a function f from the space $S_1^0(\Delta)$ that interpolates all the data in P. So, by means of an approximation scheme, we must determine the unknown coefficients $c_1, \ldots, c_n$ of f such that the resulting surface triangulation defined over Δ is "close" to the data in P, that is,

$$f(x_k, y_k) \approx z_k, \quad k = 1, \ldots, m.$$

Throughout this chapter, we will use the *method of least squares* (and variants thereof) to find f, which implies that the unknown coefficients are determined by minimizing the sum of the squared vertical distances between f and the points (x_k, y_k, z_k) for $k = 1, \ldots, m$. This implies that the quantity to be minimized can be expressed by the functional

$$I(\mathbf{c}) = \sum_{k=1}^{m} \left(f(x_k, y_k) - z_k \right)^2. \tag{8.10}$$

By inserting from (8.3) we get

$$I(\mathbf{c}) = \sum_{k=1}^{m} \left(\sum_{j=1}^{n} c_j N_j(x_k, y_k) - z_k \right)^2 = \|\mathbf{Bc} - \mathbf{z}\|_2^2. \tag{8.11}$$

In the matrix form on the right-hand side, $\mathbf{z} = (z_1, \ldots, z_m)^T$, $\mathbf{B}$ is the $m \times n$ matrix

$$\mathbf{B} = \begin{pmatrix} N_1(x_1, y_1) & N_2(x_1, y_1) & \cdots & N_n(x_1, y_1) \\ N_1(x_2, y_2) & N_2(x_2, y_2) & \cdots & N_n(x_2, y_2) \\ \vdots & \vdots & \ddots & \vdots \\ N_1(x_m, y_m) & N_2(x_m, y_m) & \cdots & N_n(x_m, y_m) \end{pmatrix}, \tag{8.12}$$

and $||\cdot||_2$ is the usual Euclidean norm where for any vector $\mathbf{u} = (u_1, \ldots, u_m)^T$,

$$||\mathbf{u}||_2 = \left(u_1^2 + \cdots + u_m^2\right)^{1/2}.$$

A minimum of the functional $I(\mathbf{c})$ occurs at a coefficient vector $\mathbf{c}^\star$ where all partial derivatives $\partial I / \partial c_i$, $i = 1, \ldots, n$ are zero. Differentiating and rearranging the subsequent expression goes as follows:

$$\frac{\partial I}{\partial c_i} = 2 \sum_{k=1}^{m} \left(\sum_{j=1}^{n} c_j N_j(x_k, y_k) - z_k \right) N_i(x_k, y_k) = 0, \quad i = 1, \ldots, n$$

$$\sum_{j=1}^{n} \sum_{k=1}^{m} N_i(x_k, y_k) N_j(x_k, y_k) c_j = \sum_{k=1}^{m} N_i(x_k, y_k) z_k, \quad i = 1, \ldots, n.$$

Thus, each differentiation gives rise to an equation in the n unknowns c_i, $i = 1, \ldots, n$ and there are exactly n equations that constitute a linear system of equations. In matrix form the equation system becomes

$$\left(\mathbf{B}^T \mathbf{B}\right) \mathbf{c} = \mathbf{B}^T \mathbf{z}, \qquad (8.13)$$

where matrix $\mathbf{B}$ and vectors $\mathbf{c}$ and $\mathbf{z}$ are given above. So, an element of the system matrix is

$$\left(\mathbf{B}^T \mathbf{B}\right)_{ij} = \sum_{k=1}^{m} N_i(x_k, y_k) N_j(x_k, y_k)$$

and an element of the right-hand side vector is

$$\left(\mathbf{B}^T \mathbf{z}\right)_i = \sum_{k=1}^{m} N_i(x_k, y_k) z_k.$$

The equations in (8.13) are called the *normal equations*. In the next section we show that the system of equations always has a unique solution.

If $m = n$, that is, if the vertices of the underlying triangulation Δ occupies all the points of P, then (8.13) reduces to $\mathbf{Bc} = \mathbf{z}$ and the minimum of $I(\mathbf{c})$ is zero. But this is just the trivial solution $\mathbf{c}^\star = \mathbf{z}$, which is the interpolating surface with $f(x_k, y_k) = z_k$ for $k = 1, \ldots, m$.

Figure 8.4 shows scattered data randomly sampled from a bivariate test function on the left, and a surface approximation to the scattered data on the right. The test function is the so-called Franke's function ([32]), which is also used in examples later,

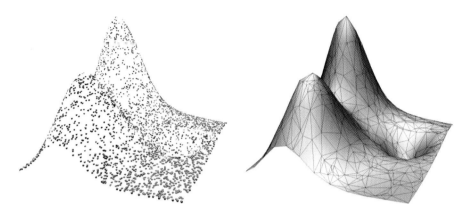

Fig. 8.4. Least squares approximation to 4500 scattered data points sampled from Franke's test function. The triangulation has approximately 500 nodes. Triangles are Gouraud-shaded.

$$f(x,y) = \frac{3}{4}e^{-\frac{(9x-2)^2+(9y-2)^2}{4}} + \frac{3}{4}e^{-\frac{(9x+1)^2}{49}-\frac{9y+1}{10}} +$$
$$\frac{1}{2}e^{-\frac{(9x-7)^2+(9y-3)^2}{4}} - \frac{1}{5}e^{-(9x-4)^2-(9y-7)^2}. \tag{8.14}$$

8.3 Existence and Uniqueness

From basic linear algebra, we know that the system of normal equations in (8.13) has at least one solution. For practical purposes, one must ensure that there is only one solution – a *unique* solution $\mathbf{c}^\star$ which gives a unique surface triangulation f as an approximation to the input data. We will use a few more results from linear algebra as a tool to arrive at uniqueness of the solution without deriving its underlying mathematical details. (See, for example, the book [33] which covers the necessary basics from linear algebra).

System (8.13) has a unique solution if the system matrix $\mathbf{B}^T\mathbf{B}$ is *positive definite*. An $n \times n$ matrix $\mathbf{A}$ is called positive definite if $\mathbf{x}^T\mathbf{A}\mathbf{x} > 0$ for all non-zero vectors $x \in R^n$. It can be shown that $\mathbf{B}^T\mathbf{B}$ is positive definite if and only if $\mathbf{B}$ has *linearly independent* columns. The columns $\mathbf{b}_1, \mathbf{b}_2, \ldots, \mathbf{b}_n$ are called linearly independent if $t_1\mathbf{b}_1 + \cdots + t_n\mathbf{b}_n = \mathbf{0}$ for some real values $\{t_j\}$ implies that $t_j = 0$ for $j = 1, \ldots, n$. Equivalently, the columns are linearly independent if it is impossible to express any one of the columns as a linear combination of the others. (The interested reader may derive the important relation between positive definiteness and linear independence in Exercise 4.)

Let $\mathbf{B}_1$ denote the $n \times n$ sub-matrix in the upper part of $\mathbf{B}$, where n is the number of vertices in the triangulation, and let $\mathbf{B}_2$ be the $(m-n) \times n$ sub-matrix in the lower part of $\mathbf{B}$. Recall the property $N_i(x_j, y_j) = \delta_{ij}$ for $j =$

$1, \ldots, n$, and the numbering (8.9) of the vertices in V. This ensures that $\mathbf{B}_1$ is the identity matrix, with 1's along the diagonal and 0's in all other positions, and $\mathbf{B}$ can be partitioned as follows,

$$
\mathbf{B} = \begin{pmatrix} \mathbf{B}_1 \\ \mathbf{B}_2 \end{pmatrix} = \left(\begin{array}{cccc}
1 & 0 & \cdots & 0 \\
0 & 1 & \cdots & 0 \\
\vdots & \vdots & \ddots & \vdots \\
0 & 0 & \cdots & 1 \\
\hline
N_1(x_{n+1}, y_{n+1}) & N_2(x_{n+1}, y_{n+1}) & \cdots & N_n(x_{n+1}, y_{n+1}) \\
N_1(x_{n+2}, y_{n+2}) & N_2(x_{n+2}, y_{n+2}) & \cdots & N_n(x_{n+2}, y_{n+2}) \\
\vdots & \vdots & \ddots & \vdots \\
N_1(x_m, y_m) & N_2(x_m, y_m) & \cdots & N_n(x_m, y_m)
\end{array} \right).
$$

$$(8.15)$$

The columns $\mathbf{b}_1^1, \ldots, \mathbf{b}_n^1$ of the identity matrix $\mathbf{B}_1$ are trivially linearly independent, so $t_1 \mathbf{b}_1^1 + \cdots + t_n \mathbf{b}_n^1 = \mathbf{0}$ implies that $t_j = 0$ for $j = 1, \ldots, n$. But then $t_1 \mathbf{b}_1 + \cdots + t_n \mathbf{b}_n = \mathbf{0}$ also implies that $t_j = 0$ for $j = 1, \ldots, n$, and the columns of $\mathbf{B}$ must also be linearly independent. This proves that the system matrix $\mathbf{B}^T \mathbf{B}$ is positive definite. Therefore the system of normal equations (8.13) has only one solution that uniquely determines a coefficient vector $\mathbf{c}^\star$ which minimizes the functional $I(\mathbf{c})$ in (8.11).[1]

Note that the analysis above relies on the fact that the vertices of the triangulation constitute a subset of the sites P of the data. In Section 8.7 we leave this restriction and approximate scattered data over triangulations with arbitrary locations of vertices.

8.4 Sparsity and Symmetry

The system matrix $\mathbf{B}^T \mathbf{B}$ of Equation (8.13) is clearly sparse and symmetric. Indeed, an element

$$
\left(\mathbf{B}^T \mathbf{B} \right)_{ij} = \sum_{k=1}^{m} N_i(x_k, y_k) N_j(x_k, y_k), \qquad (8.16)
$$

can be non-zero only when $i = j$ (on the diagonal), or when the interior of the overlap $\Omega_i \cap \Omega_j$ between the domains of the basis functions N_i and N_j is non-empty, which happens when two triangles share a common edge between the vertices v_i and v_j, see Figure 8.5. A diagonal element is always greater

---

[1] Uniqueness of the solution can also be proved by considering a lower bound for the minimal singular value $\sigma_{\min}(\mathbf{B})$ of $\mathbf{B}$, that is, the minimal eigenvalue of $(\mathbf{B}^T \mathbf{B})^{1/2}$. It can be shown that $\sigma_{\min}(\mathbf{B}) \geq 1$, which is a sufficient condition for the system matrix $\mathbf{B}^T \mathbf{B}$ to be nonsingular [70].

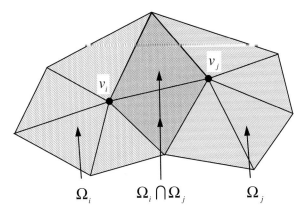

Fig. 8.5. A possible non-zero off-diagonal element in the system matrix $(B^T B)_{ij}$ corresponds to an edge between the vertices v_i and v_j in the triangulation. It is non-zero if one or more data points fall strictly inside $\Omega_i \cap \Omega_j$.

than zero, but in the latter case when $\Omega_i \cap \Omega_j \neq \phi$, then $\left(\mathbf{B}^T\mathbf{B}\right)_{ij}$ is zero if no points from P fall strictly inside $\Omega_i \cap \Omega_j$. It is also clear that all non-zero elements of $\mathbf{B}^T\mathbf{B}$ are positive. Hereafter we always say that $\left(\mathbf{B}^T\mathbf{B}\right)_{ij}$ is a non-zero element if $\Omega_i \cap \Omega_j \neq \phi$.

So, the two equal non-zero off-diagonal elements $(\mathbf{B}^T\mathbf{B})_{ij}$ and $(\mathbf{B}^T\mathbf{B})_{ji}$ correspond to an edge in the triangulation connecting the two vertices v_i and v_j. Recall from (1.7) that the number of edges in a triangulation with $|V|$ vertices has an upper bound $3|V|-6$. Then the number of non-zero off-diagonal elements in the $n \times n$ matrix $\mathbf{B}^T\mathbf{B}$ is less than $2(3|n| - 6)$, and counting the diagonal which is also non-zero, we find that there are less than $7|n| - 12$ non-zeros. Thus, the average number of non-zeros in each row of $\mathbf{B}^T\mathbf{B}$ is approximately 7. So when the number of vertices $|V|$ $(= n)$ in the underlying triangulation is large, the system matrix is extremely sparse. This is of vital importance when choosing a proper equation solver. The sparsity of the system matrix suggests that the system can be solved by an effective iterative solver like the Conjugate gradient method [42], and sparsity and symmetry can be exploited by storing $\mathbf{B}^T\mathbf{B}$ in a compact structure.

Figure 8.6 shows an example of a sparsity pattern of a 500×500 system matrix $\mathbf{B}^T\mathbf{B}$, where the non-zeros are marked with black dots. Note the high density of non-zeros around the diagonal in this example. This is due to the fact that the vertices of V in (8.9), which are vertices in the triangulation, were sorted lexicographically on their x and y value before the triangulation was made. Since a non-zero off-diagonal element in position (i,j) represents an edge between vertices v_i and v_j, it is more likely that $|i-j|$ is small after the sorting, which brings non-zero elements closer to the diagonal of the matrix.

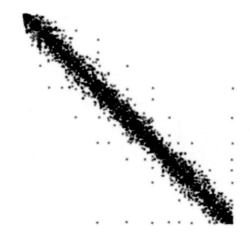

Fig. 8.6. Sparsity pattern of a 500×500 system matrix $B^T B$ for the least squares problem.

8.5 Penalized Least Squares

The resulting surface from the basic least squares approximation in Section 8.2 may not be sufficiently smooth, especially if the given data are subject to noise. This is a common case for measured data. Figure 8.7 shows two surface approximations of scattered data randomly sampled from Franke's test function. The (x, y)-positions of the data points were the same in the left and right surface, but random noise was added to the z-values of the data points on the right.

The functional $I(\mathbf{c})$ to be minimized can be augmented by a smoothing term, also called *penalty term* by Golitschek and Schumaker [85], which involves the coefficient vector $\mathbf{c}$. Many commonly used smoothing terms, including those defined below, can be written on the quadratic form

$$\mathcal{J}(\mathbf{c}) = \mathbf{c}^T \mathbf{E} \mathbf{c}, \tag{8.17}$$

where $\mathbf{E}$ is a symmetric and positive semidefinite $n \times n$ matrix. An $n \times n$ matrix $\mathbf{A}$ is called *positive semidefinite* if $\mathbf{x}^T \mathbf{A} \mathbf{x} \geq 0$ for all non-zero vectors $x \in R^n$. Recall from Section 8.3 that positive definiteness requires strict inequality, $\mathbf{x}^T \mathbf{A} \mathbf{x} > 0$. Sometimes we say that a matrix is *strictly* positive definite to distinguish it from being "only" positive semidefinite. We also say that a matrix is "at least" positive semidefinite, which means that positive semidefiniteness is guaranteed, but that positive definiteness has not yet been examined.

When working with spaces of functions of degree three and higher, smoothing terms on this form usually involve first and second derivatives. For the piecewise linear space $S_1^0(\Delta)$, which consists of functions that are not twice

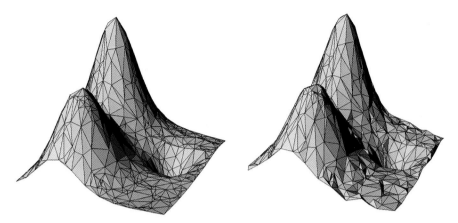

Fig. 8.7. Least squares approximation to 4500 scattered data points sampled from Franke's test function. Random noise was added to the z-values of the data points when generating the surface on the right.

differentiable, we will define discrete analogs where $\mathcal{J}(\mathbf{c})$ is the sum of some roughness measure around each vertex or across each edge of the triangulation. The functional to be minimized is now

$$I(\mathbf{c}) = \sum_{k=1}^{m} \left(f(x_k, y_k) - z_k\right)^2 + \lambda \mathcal{J}(\mathbf{c}) = \sum_{k=1}^{m} \left(f(x_k, y_k) - z_k\right)^2 + \lambda \mathbf{c}^T \mathbf{E} \mathbf{c}$$
$$= \|\mathbf{B}\mathbf{c} - \mathbf{z}\|_2^2 + \lambda \mathbf{c}^T \mathbf{E} \mathbf{c}$$

for some real value $\lambda \geq 0$, and with $\mathbf{z}$ and $\mathbf{B}$ as in Section 8.2. A minimum of $I(\mathbf{c})$ occurs where all partial derivatives $\partial I / \partial c_i$, $i = 1, \ldots, n$ are zero. Differentiating $I(\mathbf{c})$ and rearranging, we obtain the following system of normal equations,

$$\left(\mathbf{B}^T \mathbf{B} + \lambda \mathbf{E}\right) \mathbf{c} = \mathbf{B}^T \mathbf{z}, \tag{8.18}$$

with the unknown coefficient vector $\mathbf{c}$. The right-hand side is the same as in the system (8.13) for the basic least squares problem without a smoothing term, but the system matrix is augmented by $\lambda \mathbf{E}$. Recall that an equation system has a unique solution when the system matrix is positive definite. Since $\mathbf{B}^T \mathbf{B}$ is positive definite and $\mathbf{E}$ is positive semidefinite by hypothesis, it is easy to show that $\left(\mathbf{B}^T \mathbf{B} + \lambda \mathbf{E}\right)$ is positive definite (Exercise 6). Thus, (8.18) has always a unique solution under the assumption that $\mathbf{E}$ is positive semidefinite.

A critical task in practical implementations is to select the degree of smoothing through the smoothing parameter λ. A method called generalized cross-validation was proposed in [85] for computing λ, but this is rather costly. A simple and pragmatic approach can be adopted from [30] where the default value is set to

$$\lambda_d = \left\| \mathbf{B}^T \mathbf{B} \right\|_F / \left\| \mathbf{E} \right\|_F . \tag{8.19}$$

Here $\| \cdot \|_F$ denotes the Frobenius matrix norm with $\|\mathbf{A}\|_F = (\sum_{ij} a_{ij}^2)^{1/2}$ for a matrix $\mathbf{A}$. The idea is that the contribution from $\mathbf{B}^T \mathbf{B}$ and $\lambda_d \mathbf{E}$ to the system matrix should have roughly the same weight. But experiments show that when noise is significant, the smoothing parameter must be set much higher.

8.6 Smoothing Terms for Penalized Least Squares

In this section, we construct three different smoothing terms on the quadratic form in (8.17) that can be used by penalized least squares approximation defined above. For a function $g(x, y)$ that is twice differentiable, the *membrane energy*

$$\int |\nabla g|^2 = \int \left[\left(\frac{\partial g}{\partial x} \right)^2 + \left(\frac{\partial g}{\partial y} \right)^2 \right] \tag{8.20}$$

and the *thin-plate energy*

$$\int \left[\left(\frac{\partial^2 g}{\partial x^2} \right)^2 + 2 \left(\frac{\partial^2 g}{\partial x \partial y} \right)^2 + \left(\frac{\partial^2 g}{\partial y^2} \right)^2 \right] \tag{8.21}$$

can be used as smoothing terms to obtain the functional $\mathcal{J}(\mathbf{c})$ on the quadratic form in (8.17). Loosely speaking, since these terms are included in functionals to be minimized, the former prefers surfaces with small area, while the latter prefers surfaces with small overall curvature[2]. Another useful feature observed when using the thin-plate energy term is that it provides a natural extrapolation of the surface beyond the original definition area and to regions not covered with input data. This is illustrated in the example section at the end of this chapter.

Note that the thin-plate energy is based on second derivatives. Therefore the thin-plate energy cannot be used directly for the space $S_1^0(\Delta)$ of piecewise linear surface triangulations, which are not twice differentiable. A discrete analog must be constructed which resembles this energy functional. The membrane energy is based on first derivatives only and can therefore be applied directly with piecewise linear surface triangulations. Below, a discrete counterpart to the membrane energy, called the umbrella-operator, is also given. We first derive an explicit formula for the elements of the smoothing matrix $\mathbf{E}$ based on the membrane energy.

---

[2] The second derivative $h''(x)$ can be regarded as an approximation of the curvature $\kappa(x) = h''(x)/[1 + (h'(x))^2]^{3/2}$ for a univariate function $h(x)$. Thus, curvature and second derivative coincide where $h'(x) = 0$.

Membrane energy functional. Let $g_k = f|_{t_k}$ be the restriction of the piecewise linear function f to the triangle t_k. The partial derivatives for constructing the membrane energy given by (8.20) were derived in Section 8.1 and expressed by the gradient as linear combinations of the coefficients $c_1, \ldots, c_n$. To express the gradient over the triangle t_k we include the index k in Equation (8.8) and get

$$\nabla g_k = \left(\frac{\partial g_k}{\partial x}, \frac{\partial g_k}{\partial y} \right) = \frac{1}{2A_k} \left(\sum_{i=1}^{n} \eta_i^k c_i, \sum_{i=1}^{n} \mu_i^k c_i \right).$$

Let $|T|$ be the number of triangles in the triangulation Δ. The membrane energy can now be expressed on quadratic form as a sum over all triangles in Δ,

$$
\begin{aligned}
\mathcal{J}_1(\mathbf{c}) &= \sum_{k=1}^{|T|} A_k \left| \nabla g_k \right|^2 = \sum_{k=1}^{|T|} A_k \left[\left(\frac{\partial g_k}{\partial x} \right)^2 + \left(\frac{\partial g_k}{\partial y} \right)^2 \right] \\
&= \sum_{k=1}^{|T|} \frac{1}{4A_k} \left[\left(\sum_{i=1}^{n} \eta_i^k c_i \right)^2 + \left(\sum_{i=1}^{n} \mu_i^k c_i \right)^2 \right] \\
&= \sum_{k=1}^{|T|} \frac{1}{4A_k} \left[\left(\sum_{i=1}^{n} \eta_i^k c_i \right) \left(\sum_{j=1}^{n} \eta_j^k c_j \right) + \left(\sum_{i=1}^{n} \mu_i^k c_i \right) \left(\sum_{j=1}^{n} \mu_j^k c_j \right) \right] \\
&= \sum_{i=1}^{n} \sum_{i=j}^{n} \left(\sum_{k=1}^{|T|} \left(\eta_i^k \eta_j^k + \mu_i^k \mu_j^k \right) / 4A_k \right) c_i c_j = \mathbf{c}^T \mathbf{E} \mathbf{c},
\end{aligned}
$$

where

$$E_{ij} = \sum_{k=1}^{|T|} (\eta_i^k \eta_j^k + \mu_i^k \mu_j^k)/4A_k.$$

Matrix $\mathbf{E}$ is clearly symmetric, and it is also sparse. Indeed an element E_{ij} can be non-zero only when i and j are vertex indices of the same triangle (cf. Section 8.1). Thus a matrix element E_{ij} can be non-zero only when (i, j) corresponds to an edge in the triangulation connecting the two vertices v_i and v_j. This is exactly the same sparsity pattern as that of the system matrix $\mathbf{B}^T \mathbf{B}$ of the basic least squares problem without a smoothing term.

As stated in Section 8.2, a unique solution of (8.18) relies on the condition that $\mathbf{E}$ is (at least) positive semidefinite. When deriving $\mathcal{J}_1(\mathbf{c}) = \mathbf{c}^T \mathbf{E} \mathbf{c}$ above, it is clear from the intermediate results that $\mathbf{c}^T \mathbf{E} \mathbf{c}$ can never be less than zero. Thus, $\mathbf{E}$ is positive semidefinite and the penalized least squares problem has a unique solution when $\mathbf{E}$ is deduced from the membrane energy.

The Umbrella-operator. Here we derive a discrete version of the membrane energy functional. It is included here because of its relatively simple formulation and as an example of discretization. In practical implementations, the continuous counterpart above and the discrete version of the thin-plate energy derived below are more useful.

Let $\partial\Omega_k$ denote the set of indices of the vertices on the boundary of the support Ω_k of the basis function $N_k(x,y)$ depicted in Figure 8.1. Further, let n_k denote the degree (or valency) of the vertex v_k. We remember from Section 1.3 that the degree of v_k is the number of edges that meet with v_k as a common vertex. Then, for each vertex in Δ we make a discrete roughness measure around v_k as a linear combination of the coefficient associated with v_k and the neighborhood of vertices adjacent to v_k,

$$\mathcal{M}_k f = \frac{1}{n_k} \sum_{l \in \partial\Omega_k} c_l - c_k.$$

This is called the *umbrella-operator* by some authors, for example in [48] where it is used as a smoothing operator in subdivision schemes for multiresolution modeling. See also the umbrella Figure in 8.8. Here we will regard the squared expression $(\mathcal{M}_k f)^2$ as a discrete analog of the integrand in the functional (8.20). One may argue that the umbrella-operator does not make sense when c_k corresponds to a vertex at the boundary of the triangulation. Experiments also show that it may produce artifacts both at the boundary and interior to the mesh.

If $\mathcal{M}_k f$ is expressed as a linear combination of all the coefficients in the unknown coefficient vector $\mathbf{c}$, we obtain

$$\mathcal{M}_k f = \sum_{l=1}^{n} \rho_l^k c_l, \quad \rho_l^k = \begin{cases} -1, \, l = k \\ \frac{1}{n_k}, \text{ if } (v_k, v_l) \text{ is an edge in } \Delta \\ 0 \quad \text{otherwise.} \end{cases} \tag{8.22}$$

Applying the umbrella-operator to all vertices in the triangulation and collecting the contributions, the membrane energy representing the smoothing term can be expressed on the quadratic form

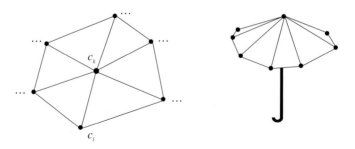

Fig. 8.8. The umbrella-operator.

$$\mathcal{J}_2(\mathbf{c}) = \sum_{k=1}^{n} (\mathcal{M}_k f)^2 = \sum_{k=1}^{n} \left[\sum_{i=1}^{n} \rho_i^k c_i \right]^2$$

$$= \sum_{k=1}^{n} \left[\sum_{i=1}^{n} \rho_i^k c_i \right] \left[\sum_{j=1}^{n} \rho_j^k c_j \right]$$

$$= \sum_{i=1}^{n} \sum_{j=1}^{n} \left[\sum_{k=1}^{n} \rho_i^k \rho_j^k \right] c_i c_j = \sum_{i=1}^{n} \sum_{j=1}^{n} E_{ij} c_i c_j = \mathbf{c}^T \mathbf{E} \mathbf{c}, \qquad (8.23)$$

where

$$E_{ij} = \sum_{k=1}^{n} \rho_i^k \rho_j^k.$$

Matrix $\mathbf{E}$ is clearly symmetric, which also gives a symmetric system matrix $\left(\mathbf{B}^T \mathbf{B} + \lambda \mathbf{E} \right)$ since $\mathbf{B}^T \mathbf{B}$ from the basic least squares problem is symmetric. It is also sparse, but the number of non-zero elements in $\mathbf{E}$ is larger than in $\mathbf{B}^T \mathbf{B}$. As for $(\mathbf{B}^T \mathbf{B})_{ij}$, E_{ij} is non-zero when $i = j$ and when (i, j) corresponds to an edge in the triangulation connecting the two vertices v_i and v_j. In addition, we see that $\rho_i^k \rho_j^k$ (and thus E_{ij}) is non-zero if both (i, k) and (j, k) correspond to edges in the triangulation. In Figure 8.9 this is illustrated by the vertices marked '2' and '3' that generate non-zero entries in combination with vertex v_i.

A straightforward extension of the umbrella-operator is to incorporate weights, with weights being inverse edge lengths. We get the weighted operator

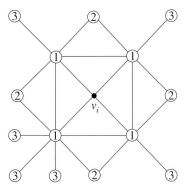

Fig. 8.9. Illustration of which triangle vertices that generate non-zero off-diagonal elements in the system matrix together with the vertex v_i. The '1'-vertices generate non-zeros with v_i by the membrane energy and in the basic least squares problem (matrix $\mathbf{B}^T \mathbf{B}$); the '1' and '2'-vertices generate non-zeros by the thin-plate energy term; and '1', '2' and '3'-vertices generate non-zeros by the umbrella-operator.

$$\widetilde{\mathcal{M}}_k = \frac{1}{W_k} \sum_{l \in \partial \Omega_k} \omega_{kl} c_l - c_k,$$

where $\omega_{kl} = 1/\|v_l - v_k\|_2$ and $W_k = \sum_{l \in \partial \Omega_k} \omega_{kl}$. This weighting makes the operator scale dependent and may help to avoid distortion when edge lengths vary over the triangulation. The energy functional takes the same quadratic form with the same sparsity pattern as the unweighted version but now with coefficients $\widetilde{\rho}_l^k$ replacing ρ_l^k in (8.22) where

$$\widetilde{\rho}_l^k = \begin{cases} -1, \, l = k \\ \frac{\omega_{kl}}{W_k}, \, \text{if } (v_k, v_l) \text{ is an edge in } \Delta \\ 0 \quad \text{otherwise.} \end{cases}$$

In the special case where all edge lengths are equal, we get the same roughness measure as in the unweighted version.

As for the first membrane energy measure above, it is clear from the intermediate results in (8.23) that $\mathbf{c}^T \mathbf{E} \mathbf{c}$ can never be less than zero. Thus, $\mathbf{E}$ is positive semidefinite and the penalized least squares problem has a unique solution when $\mathbf{E}$ is constructed by the umbrella-operator.

Discrete thin-plate energy functional. A discrete analog of the thin-plate energy functional (8.21) can be constructed from a second order divided difference. Figure 8.10 shows a stencil with the triangles and vertices involved in the construction of the divided difference operator below. Let $g_i = f|_{t_i}$ for $i = 1, 2$ be restrictions of the piecewise linear function f to the triangles t_1 and t_2 sharing the edge e_k. Further, let $\mathbf{n}_{e_k}$ be a unit vector in the xy-plane orthogonal to the projection of e_k in the xy-plane with direction as shown in Figure 8.10. The derivatives of g_1 and g_2 in the direction of $\mathbf{n}_{e_k}$ are

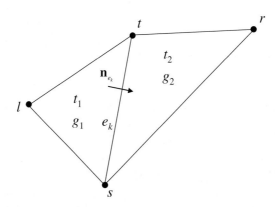

Fig. 8.10. Stencil for the second order divided difference $D_k^2 f$ used to make the discrete thin-plate energy measure.

$$\frac{\partial g_i}{\partial \mathbf{n}_{e_k}} = \left(\frac{\partial g_i}{\partial x}, \frac{\partial g_i}{\partial y}\right) \cdot \mathbf{n}_{e_k} = \nabla g_i \cdot \mathbf{n}_{e_k}, \quad i = 1, 2.$$

We define the second order divided difference associated with an interior edge e_k as the jump in the derivative of f over e_k in the direction of $\mathbf{n}_{e_k}$,

$$\begin{aligned}
\mathcal{T}_k f &= \frac{\partial g_2}{\partial \mathbf{n}_{e_k}} - \frac{\partial g_1}{\partial \mathbf{n}_{e_k}} \\
&= \left(\frac{\partial g_2}{\partial x} - \frac{\partial g_1}{\partial x}, \frac{\partial g_2}{\partial y} - \frac{\partial g_1}{\partial y}\right) \cdot \mathbf{n}_{e_k} \\
&= (\nabla g_2 - \nabla g_1) \cdot \mathbf{n}_{e_k}
\end{aligned} \tag{8.24}$$

This is exactly the same measure as used by the data dependent swapping criterion called *jump in normal derivative* (JND) in Section 5.4. Equivalent measures have also been used as divided difference operators in subdivision schemes for triangulations, see for example [36] and [37].

To ease implementation, we give an explicit formula for $\mathcal{T}_k f$ expressed by the coefficients associated with the vertices of the stencil in Figure 8.10 and the 2D coordinates of the vertices of the stencil. Let $\omega(e_k) = (s, t, l, r)$ be the indices of the four vertices (v_s, v_t, v_l, v_r) in the neighborhood of an interior edge e_k. Let v_s and v_t be the endpoints of e_k, and let v_l and v_r be the vertices to the left and to the right of e_k respectively as seen from v_s towards v_t, see Figure 8.10. Further, let A_{abc} be the area of a triangle in the xy-plane spanned by vertices with indices a, b and c. The area can be found by Equation (8.6).

The gradients over the triangles t_1 and t_2 of the stencil can be found by using Equation (8.7). We include indices of the triangles t_1 and t_2 and get

$$\begin{aligned}
\nabla g_1 &= \left(\eta_l c_l + \eta_s^1 c_s + \eta_t^1 c_t, \ \mu_l c_l + \mu_s^1 c_s + \mu_t^1 c_t\right)/2A_{lst}, \\
\nabla g_2 &= \left(\eta_r c_r + \eta_t^2 c_t + \eta_s^2 c_s, \ \mu_r c_r + \mu_t^2 c_t + \mu_s^2 c_s\right)/2A_{rts}.
\end{aligned} \tag{8.25}$$

We also get

$$\mathbf{n}_{e_k} = (y_t - y_s, x_s - x_t)/L_{e_k}, \tag{8.26}$$

where $L_{e_k} = \sqrt{(x_t - x_s)^2 + (y_t - y_s)^2}$ is the length of the projection of e_k onto the xy-plane.

From (8.25) we see that $\mathcal{T}_k f = (\nabla g_2 - \nabla g_1) \cdot \mathbf{n}_{e_k}$ is a linear combination of the coefficients (c_s, c_t, c_l, c_r). Leaving out details of the calculation here (see Exercise 7), we arrive at these formulas when we insert (8.25) and (8.26) into (8.24),

$$\mathcal{T}_k f = \sum_{i \in \omega(e_k)} \beta_i^k c_i, \tag{8.27}$$

where

$$\beta_l^k = -\frac{L_{e_k}}{2A_{lst}} \qquad \beta_r^k = -\frac{L_{e_k}}{2A_{rts}}$$

$$\beta_s^k = \frac{L_{e_k} A_{tlr}}{2A_{lst} A_{rts}} \qquad \beta_t^k = \frac{L_{e_k} A_{srl}}{2A_{lst} A_{rts}}.$$

Let $|E_I|$ denote the number of interior edges in Δ. We express the discrete thin-plate energy as a sum over all interior edges. Since $\mathcal{T}_k f$ is linear, it generates a quadratic energy term, which was also the case for the membrane energy functional. We obtain the thin-plate energy functional

$$\mathcal{J}_3(\mathbf{c}) = \sum_{k=1}^{|E_I|} L_{e_k} \left(\mathcal{T}_k f\right)^2 = \sum_{k=1}^{|E_I|} L_{e_k} \left[\sum_{i=1}^{n} \beta_i^k c_i\right]^2$$

$$= \sum_{k=1}^{|E_I|} L_{e_k} \left[\sum_{i=1}^{n} \beta_i^k c_i\right] \left[\sum_{j=1}^{n} \beta_j^k c_j\right]$$

$$= \sum_{i=1}^{n} \sum_{j=1}^{n} \left[\sum_{k=1}^{|E_I|} L_{e_k} \beta_i^k \beta_j^k\right] c_i c_j = \sum_{i=1}^{n} \sum_{j=1}^{n} E_{ij} c_i c_j = \mathbf{c}^T \mathbf{E} \mathbf{c},$$

where

$$E_{ij} = \sum_{k=1}^{|E_I|} L_{e_k} \beta_i^k \beta_j^k.$$

In general, an element E_{ij} is non-zero if $i = j$, or if both i and j correspond to vertices of the same neighborhood $\omega(e_k)$ of at least one edge e_k in the triangulation. This occurs if the line segment (v_i, v_j) spans an edge, or if v_i is the vertex on the opposite side of an edge from v_j. The latter case corresponds to the vertices with indices l and r of $\omega(e_k)$ in Figure 8.10, which give a non-zero matrix element E_{lr}.

Compared to the membrane energy, the thin-plate energy generates a sparsity pattern with more non-zero entries in the system matrix, but compared to the umbrella-operator, the number of non-zeros is fewer, see Figure 8.9.

Using the same arguments as above for the membrane energy functional, we find that $\mathbf{E}$ is also positive semidefinite when derived from the discrete thin-plate energy, and consequently the penalized least squares problem in (8.18) has a unique solution.

Figure 8.11 shows a surface approximation of the same noisy point set that was used by the approximation in the right plot of Figure 8.7, but now with a smoothing term constructed from the thin-plate energy functional. The smoothing parameter λ was set to $(100 \cdot \|\mathbf{B}^T \mathbf{B}\|_F / \|\mathbf{E}\|_F)$, where $\|\cdot\|_F$ denotes the Frobenius matrix norm, see Section 8.5.

To justify the use of the smoothing term for controlling the behavior of the least square fit, we briefly go through a result by Golitschek & Schumaker [85].

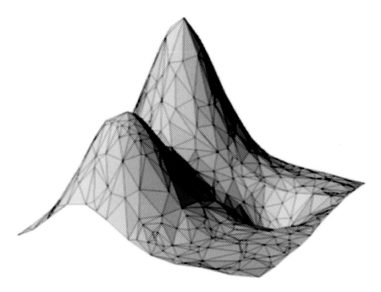

Fig. 8.11. Penalized least squares aproximation of a noisy point set.

Let $f_\lambda(x, y)$ be the penalized least squares fit to the given data with smoothing parameter λ. Then the mean square error given by

$$T_{\mathbf{z}}(\lambda) = \frac{1}{m} \sum_{k=1}^{m} \left(f_\lambda(x_k, y_k) - z_k\right)^2 \tag{8.28}$$

is monotone increasing for $\lambda \geq 0$ with derivatives $T_{\mathbf{z}}'(0) = 0$ and $\lim_{\lambda \to \infty} T_{\mathbf{z}}'(\lambda) = 0$. Thus, λ controls the trade-off between smoothness and mean square error of the fit. A best fit, best in the sense of minimizing the mean square error, is then obtained by setting the value of the smoothing parameter λ to zero.

8.7 Approximation over General Triangulations

The analysis of uniqueness in Section 8.3 of the basic least squares problem without a smoothing term, relied on the restriction that all vertices of the underlying triangulation Δ were chosen among the sites P of the input data. This condition produced the matrix $\mathbf{B}$ in (8.15) with linear independent columns. Consequently, the system matrix $\mathbf{B}^T\mathbf{B}$ was positive definite, which was a sufficient condition for the equation system in (8.13) to have a unique solution.

In this section, the scheme is made more general by removing the restriction about vertices of Δ being a subset of the sites P of the input data. We

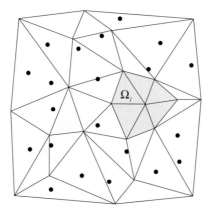

Fig. 8.12. Point set for penalized least squares. The shaded region is a domain of a basis function Ω_j that do not cover any data points.

must now revisit the general form of B in (8.12), which does not rely on this restriction.

If one or more columns of $\mathbf{B}$ are zero vectors, then the columns are not linearly independent and a unique solution of the equation system is not guaranteed since $\mathbf{B}^T\mathbf{B}$ is not positive definite in general. Note that in the matrix $\mathbf{B}$ in (8.12), each point (x_i, y_i), $i = 1, \ldots, m$ from P generates a row and each basis function N_j, $j = 1, \ldots, n$ makes a column. A column, say column number l, is a zero vector if $N_l(x_i, y_i) = 0$ for $i = 1, \ldots, m$. This happens if no point of P falls strictly inside the domain Ω_l of N_l. The situation is illustrated in Figure 8.12. This often occurs when constructing surfaces from cartographic data and geological data. For instance, contour data in cartography and seismic data (track data) in geology typically possess such uneven distributions. A non-unique solution in this case can also be understood from the functional $I(\mathbf{c})$ in (8.11) to be minimized. If there is no data point inside the domain Ω_l of a basis function N_l, then the corresponding coefficient c_l can have an arbitrary value without affecting $I(\mathbf{c})$.

So, in general $\mathbf{B}^T\mathbf{B}$ is not positive definite, but positive semidefinite for general triangulations. It is indeed positive semidefinite because

$$\mathbf{x}^T \left(\mathbf{B}^T\mathbf{B} \right) \mathbf{x} = \left(\mathbf{B}\mathbf{x} \right)^T \left(\mathbf{B}\mathbf{x} \right) = \| \mathbf{B}\mathbf{x} \|_2^2 \geq 0.$$

To compensate for the lack of positive definiteness of $\mathbf{B}^T\mathbf{B}$, we may add a smoothing term representing the membrane energy or the thin-plate energy derived in the previous section. The linear equation system $\left(\mathbf{B}^T\mathbf{B} + \lambda\mathbf{E} \right) \mathbf{c} = \mathbf{B}^T\mathbf{z}$ has a unique solution if the system matrix $\left(\mathbf{B}^T\mathbf{B} + \lambda\mathbf{E} \right)$ is positive definite. It follows from basic linear algebra that since both matrices $\mathbf{B}^T\mathbf{B}$ and $\mathbf{E}$ are positive semidefinite, then the system matrix $\left(\mathbf{B}^T\mathbf{B} + \lambda\mathbf{E} \right)$ for $\lambda > 0$ is either positive semidefinite or strictly positive definite.

For $(\mathbf{B}^T\mathbf{B} + \lambda\mathbf{E})$ to be positive semidefinite, we must have

$$\mathbf{c}^T\left(\mathbf{B}^T\mathbf{B} + \lambda\mathbf{E}\right)\mathbf{c} = \mathbf{c}^T\left(\mathbf{B}^T\mathbf{B}\right)\mathbf{c} + \lambda\mathbf{c}^T\mathbf{E}\mathbf{c} = 0, \qquad (8.29)$$

for some vector $\mathbf{c} \neq 0$. It is clear from above that $\mathbf{c}^T\left(\mathbf{B}^T\mathbf{B}\right)\mathbf{c} \geq 0$, and since $\mathbf{E}$ is positive semidefinite, we also have $\mathbf{c}^T\mathbf{E}\mathbf{c} \geq 0$. Thus, (8.29) only holds if both terms $\mathbf{c}^T(\mathbf{B}^T\mathbf{B})\mathbf{c} = 0$ and $\mathbf{c}^T\mathbf{E}\mathbf{c} = 0$. Below we analyze the system matrix $\left(\mathbf{B}^T\mathbf{B} + \lambda\mathbf{E}\right)$ in view of these conditions by geometric reasoning. In fact, we will find that $\mathbf{c}$ must be the zero vector for (8.29) to hold. Thus, the system matrix is indeed strictly positive definite when the smoothing term is constructed from the membrane energy functionals and the thin-plate energy. Consequently, the system of equations in Equation (8.18) always has a unique solution.

We first observe that $\mathbf{c}^T\mathbf{E}\mathbf{c} = \mathcal{J}(\mathbf{c})$, which is the general energy term.

Uniqueness with the membrane energy functional. The membrane energy generates the functional

$$\mathcal{J}_1(\mathbf{c}) = \sum_{k=1}^{|T|} A_k\left[\left(\frac{\partial g_k}{\partial x}\right)^2 + \left(\frac{\partial g_k}{\partial y}\right)^2\right] = \mathbf{c}^T\mathbf{E}\mathbf{c}.$$

For this expression to be 0 we must have $\partial g_k/\partial x = 0$ and $\partial g_k/\partial y = 0$ for $k = 1, \ldots, |T|$. But then, since the triangulation is connected, all coefficients must be equal. Thus, $\mathbf{c}^T\mathbf{E}\mathbf{c} = 0$ if and only if $c_1 = c_2 = \cdots = c_n$.

Furthermore, for the first term in (8.29) stemming from the basic least squares problem, we must have

$$\mathbf{c}^T(\mathbf{B}^T\mathbf{B})\mathbf{c} = (\mathbf{B}\mathbf{c})^T(\mathbf{B}\mathbf{c}) = \|\mathbf{B}\mathbf{c}\|_2^2 = 0, \qquad (8.30)$$

which implies that $(\mathbf{B}\mathbf{c})_j = \sum_{i=1}^{n} c_i N_i(x_j, y_j) = f(x_j, y_j) = 0$ for all $j = 1, \ldots, m$. But since all coefficients c_k are equal, the function f must then be zero everywhere and all coefficients c_k must be zero. Thus, $\mathbf{c}^T(\mathbf{B}^T\mathbf{B} + \lambda\mathbf{E})\mathbf{c} = 0$ implies that $\mathbf{c} = 0$ and we conclude that the system matrix $(\mathbf{B}^T\mathbf{B} + \lambda\mathbf{E})$ is indeed positive definite. Hence, the system of equations (8.18) always has a unique solution when $\mathbf{E}$ is constructed from the membrane energy.

Uniqueness with the umbrella-operator. The unweighted umbrella-operator $\mathcal{M}_k$ generates the membrane energy functional

$$\mathcal{J}_2(\mathbf{c}) = \sum_{k=1}^{n} (\mathcal{M}_k f)^2 = \sum_{k=1}^{n}\left[\frac{1}{n_k}\sum_{l \in \partial\Omega_k} c_l - c_k\right]^2.$$

For this expression to be zero we must have

$$c_k = \frac{1}{n_k} \sum_{l \in \partial \Omega_k} c_l, \quad k = 1, \ldots n. \tag{8.31}$$

In particular, this must hold for the largest coefficient $c_{max} = \max\{c_k\}_{k=1}^n$. But then all neighbors c_l of c_{max} with l in $\partial \Omega_{max}$ must be equal to c_{max}. Since the triangulation is connected, this argument can be applied repeatedly for all coefficients c_k and we find that $c^T \mathbf{E} c = 0$ if and only if $c_1 = c_2 = \cdots = c_n$. But then, (8.30) again implies that $f(x_j, y_j) = 0$ for $j = 1, \ldots, m$ under the same assumption that $\mathbf{B}^T \mathbf{B}$ is positive semidefinite. This implies, that f is zero everywhere and that all coefficients in $\mathbf{c}$ must be zero. Thus, we arrive at the same conclusion as above, that the system matrix $(\mathbf{B}^T \mathbf{B} + \lambda \mathbf{E})$ is positive definite and that (8.18) always has a unique solution also when $\mathbf{E}$ is derived from the umbrella-operator.

For the weighted umbrella-operator, we easily arrive at the same conclusion by replacing (8.31) with the convex combination

$$c_k = \frac{1}{W_k} \sum_{l \in \partial \Omega_k} \omega_{kl} c_l, \quad k = 1, \ldots n$$

and using the same arguments as above.

Uniqueness with the thin-plate energy functional. The second order difference operator $\mathcal{T}_k f$ generates the thin-plate energy functional

$$\mathcal{J}_3(\mathbf{c}) = \sum_{k=1}^{|E_I|} (\mathcal{T}_k f)^2 = \sum_{k=1}^{|E_I|} [(\nabla g_2 - \nabla g_1) \cdot \mathbf{n}_{e_k}]^2 = \mathbf{c}^T \mathbf{E} \mathbf{c}.$$

For this expression to be 0 for $\mathbf{c} \neq 0$, and thus for $\mathbf{E}$ to be positive semidefinite, we must have $(\nabla g_2 - \nabla g_1) = \mathbf{0}$ over all interior edges in the triangulation. Then, since the triangulation is connected, all the gradients of $f|_{t_i}$ over all triangles t_i in Δ are equal and f must be a linear polynomial. Again (8.30) implies that $f(x_j, y_j) = 0$ for $j = 1, \ldots, m$ under the assumption that $\mathbf{B}^T \mathbf{B}$ is positive semidefinite. From here we follow the same line of proof as above and conclude that (8.18) has a unique solution when $\mathbf{E}$ is derived from the thin-plate energy functional.

8.8 Weighted Least Squares

Assume that there is a weight of importance $w_k > 0$ (or alternatively an uncertainty $1/w_k$) associated with the z-value of each incoming data point (x_k, y_k, z_k) in the least squares formulation in Section 8.2, or in Section 8.5 when including a smoothing term. That is, we want to weight the points differently in order to give high-quality points more weight and low-quality points less weight, and thereby attract the least squares fit to certain data points with high importance. The weights can be multiplied by the squared deviations such as to modify the functional (8.10) as follows,

$$I(\mathbf{c}) = \sum_{k=1}^{m} w_k \left[f(x_k, y_k) - z_k\right]^2 = \sum_{k=1}^{m} w_k \left[\sum_{j=1}^{n} c_j N_j(x_k, y_k) - z_k\right]^2.$$

We derive the normal equations (8.13) accordingly by differentiating as in Section 8.2 and get

$$\frac{\partial I}{\partial c_i} = 2 \sum_{k=1}^{m} w_k \left[\sum_{j=1}^{n} c_j N_j(x_k, y_k) - z_k\right] N_i(x_k, y_k) = 0, \quad i = 1, \dots, n,$$

and rearrange to obtain the normal equations,

$$\sum_{j=1}^{n}\sum_{k=1}^{m} w_k N_i(x_k, y_k) N_j(x_k, y_k) c_j = \sum_{k=1}^{m} w_k N_i(x_k, y_k) z_k, \quad i = 1, \dots, n.$$

In matrix form, the normal equations can be written exactly as in Section 8.2,

$$\left(\mathbf{B}^T \mathbf{B}\right) \mathbf{c} = \mathbf{B}^T \mathbf{z},$$

but now with weights included in the matrix elements and in the elements of the right-hand side vector,

$$\left(\mathbf{B}^T \mathbf{B}\right)_{ij} = \sum_{k=1}^{m} w_k N_i(x_k, y_k) N_j(x_k, y_k), \quad i, j = 1, \dots, n, \qquad (8.32a)$$

$$\left(\mathbf{B}^T \mathbf{z}\right)_i = \sum_{k=1}^{m} w_k N_i(x_k, y_k) z_k, \quad i = 1, \dots, n. \qquad (8.32b)$$

In the special case where all weights are equal, we get the same result as in the unweighted version of the normal equations. It is straightforward to include a smoothing term for approximating noisy data (Section 8.5) or to guarantee a unique solution when creating approximations over general triangulations (Section 8.7). The equation system can be written as before in matrix form, $\left(\mathbf{B}^T \mathbf{B} + \lambda \mathbf{E}\right) \mathbf{c} = \mathbf{B}^T \mathbf{z}$, but with weights included in the matrix elements and in the elements of the right-hand side vector as above in (8.32a) and (8.32b). In practical application one will typically operate with default weights ($= 1$) for most input data, and then assign weights with values greater than one to certain data points with some prescribed high significance.

8.9 Constrained Least Squares

Suppose that the least squares approximation problem is subject to constraints $f(x_r, y_r) = z_r$ for a subset $\Gamma = \{(x_r, y_r, z_r)\}$ of the given scattered data. That is, we want the surface approximation to interpolate certain points among the input data, while the other data points are approximated by least squares under the given constraints. For example, the overall scattered data might be seabed data obtained from multi-beam echo sounders, while the constraints might be exact measurements obtained from other devices. Typically, there will be hundreds of thousands, or millions of data points from multi-beam echo sounders, but relatively few exact measurements that should be interpolated.

To ease the mathematical formulation, we assume that the locations $\{(x_r, y_r)\}$ of the constraints are among the vertices of the underlying triangulation Δ. Further, to ease the notation, we assume without loss of generality, that there are $n + \gamma$ vertices in Δ, where γ is the number of constraints, and that the numbering of the scattered data which define the constraints is as follows,

$$\Gamma = \{(x_r, y_r, z_r)\}_{r=n+1}^{n+\gamma}.$$

Thus, at least γ of the vertices in Δ are among the locations of the given scattered data, but all vertices of Δ are not necessarily among the sites of the input data as in Section 8.2. The linear combination that defines the surface triangulations takes the same form as (8.3), but now with $n+\gamma$ basis functions and corresponding coefficients,

$$f(x, y) = \sum_{j=1}^{n+\gamma} c_j N_j(x, y).$$

In order to meet the constraints, $f(x_r, y_r) = z_r$ for $r = n + 1, \ldots, n + \gamma$, the coefficients $(c_{n+1}, \ldots, c_{n+\gamma})$ must be pairwise equal to the given constraint data values $(z_{n+1}, \ldots, z_{n+\gamma})$. This follows directly from the property of the basis functions which satisfy $N_i(x_r, y_r) = \delta_{ir}$, where (x_r, y_r) is the location of a constraint and therefore the location of a vertex in Δ. Thus, the constraint data values $z_{n+1}, \ldots, z_{n+\gamma}$ can be regarded as predetermined coefficients of the surface triangulation. The linear combination can then be split into two terms, one corresponding to n unknown coefficients and one corresponding to γ predetermined coefficients,

$$f(x, y) = \sum_{j=1}^{n} c_j N_j(x, y) + \underline{\sum_{r=n+1}^{n+\gamma} z_r N_r(x, y)}.$$

The underlined expression, which is retained on the same form below when deriving the normal equations, corresponds to contributions from the constraints. Omitting the underlined expression brings us back to the same basic least squares formulation as in Section 8.2. The functional (8.10) expressing the sum of squared deviations can now be written as,

$$I(\mathbf{c}) = \sum_{k=1}^{m} \left[\sum_{j=1}^{n} c_j N_j(x_k, y_k) + \underline{\sum_{r=n+1}^{n+\gamma} z_r N_r(x_k, y_k)} - z_k \right]^2 .$$

Differentiating in each of the n unknown coefficients and setting the partial derivatives $\partial I/\partial c_i$, $i = 1, \ldots, n$ equal to zero gives

$$\frac{\partial I}{\partial c_i} = 2 \sum_{k=1}^{m} \left[\sum_{j=1}^{n} c_j N_j(x_k, y_k) + \underline{\sum_{r=n+1}^{n+\gamma} z_r N_r(x_k, y_k)} - z_k \right] N_i(x_k, y_k) = 0,$$

for $i = 1, \ldots, n$. Rearranging we get the n normal equations,

$$\sum_{j=1}^{n} \sum_{k=1}^{m} N_i(x_k, y_k) N_j(x_k, y_k) c_j = \sum_{k=1}^{m} N_i(x_k, y_k) \left[z_k - \underline{\sum_{r=n+1}^{n+\gamma} z_r N_r(x_k, y_k)} \right],$$

for $i = 1, \ldots, n$. In matrix form the normal equations can be written exactly as in Section 8.2, and as for weighted least squares in the previous section,

$$\left(\mathbf{B}^T \mathbf{B} \right) \mathbf{c} = \mathbf{B}^T \mathbf{z},$$

where

$$\left(\mathbf{B}^T \mathbf{B} \right)_{ij} = \sum_{k=1}^{m} N_i(x_k, y_k) N_j(x_k, y_k), \quad i, j = 1, \ldots, n, \quad \text{and}$$

$$\left(\mathbf{B}^T \mathbf{z} \right)_i = \sum_{k=1}^{m} N_i(x_k, y_k) \left[z_k - \underline{\sum_{r=n+1}^{n+\gamma} z_r N_r(x_k, y_k)} \right], \quad i = 1, \ldots, n.$$

As for weighted least squares, it is straightforward to include a smoothing term. Weighted least squares and constraints can also be combined (Exercise 8).

Since the surface approximation $f(x, y)$ is piecewise linear, it is linear along edges of the triangulation. Therefore, if both endpoints of an edge are among the constraints in Γ, then this edge will be embedded as a predefined straight line segment in the surface approximation. Thus, one can easily embed predefined networks in the surface approximation representing roads, river systems with lake boundaries, or geological fault networks. This is a useful feature, for example in terrain modeling to guarantee consistent drainage systems.

8.10 Approximation over Binary Triangulations

The remainder of this chapter is devoted to least squares approximation of huge scattered data sets and data that are unevenly distributed over the domain. We apply a subdivision scheme based on binary triangulations that was outlined in Section 2.10. The scheme generates an adaptive triangulation where the triangle density reflects the variation in surface topography and distribution of the given data.

One important property of this simple, though very powerful, technique is that refinement is done locally without the need to maintain the entire mesh at the same resolution. Apart from being an effective means for obtaining level-of-detail control in large-scale visualization, we demonstrate here that binary triangulations are powerful when applied in a multilevel least squares approximation scheme. The fact that binary triangulations are standard tools for view-dependent visualization also makes the resulting surfaces well suited for fast rendering. In particular, least squares approximation over binary triangulations is efficient when fitting surfaces to huge sets of data and to data that are unevenly distributed over the domain.

We start below by outlining principles of a general multilevel approximation scheme independent of which type of triangulation the surface is defined over. Then details of the least squares approximation scheme over binary triangulations are given. We refer to Section 2.10 for the basics on binary triangulations. In Section 8.11, several numerical examples demonstrate the use of the scheme.

The general multilevel scheme. Let us assume that we are given a set of non-collinear scattered data sites $P = \{(x_1, y_1), \ldots, (x_m, y_m)\}$ with corresponding data values $\{z_1, \ldots, z_m\}$. Over each triangulation Δ_k in a coarse-to-fine sequence of triangulations we construct a least squares surface approximation f_{Δ_k} to the given data. We want each triangulation in the sequence to reflect the distribution of the underlying scattered data, or depend on the quality (accuracy) of the approximation found at coarser levels. Thus, we use the approximation at one level to decide where to refine, hence to produce the triangulation at the next finer level. Since iterative equation solvers are used for computing the least squares approximation, the solution from one level can be used as an initial guess (starting vector) for the solution at the next level. In the very general Algorithm 8.1, ϵ is a prescribed tolerance. That is, ϵ is the maximum allowed deviation $|f_{\Delta}(x_i, y_i) - z_i|$, $i = 1, \ldots, m$, where Δ is the triangulation at the finest level.

Below, refinement is done by longest edge bisection and every triangulation Δ_k in the coarse-to-fine sequence is a binary triangulation. Dependent on the application, criteria for refinement other than a tolerance ϵ can be used. For huge data sets it is important to reduce the amount of data or to create a triangulation that reflects the distribution of the scattered data. One

---

Algorithm 8.1 General multilevel scheme

1. **for** $k = 1, 2, 3, \ldots$
2.   Find a surface approximation f_{Δ_k}.
3.   **if** $|f_{\Delta_k}(x_i, y_i) - z_i| < \epsilon, \ i = 1, \ldots, m,$
       stop.
4.   **else**
5.   Refine Δ_k locally where ϵ is exceeded to produce Δ_{k+1},
           and use f_{Δ_k} to make an initial guess for $f_{\Delta_{k+1}}$.

---

way to achieve this, without taking any error measure into account, is to use a counting measure whereby Δ_k is refined as long as there is more than a prescribed number of data locations from P within some portion of the domain. Variation over curvature measures could also be used to decide where to refine, for example by using the thin-plate energy measure in (8.27) locally over triangle edges and refine the triangulation where this measure exceeds a given threshold. Different measures could also be combined.

Approximation Scheme for Binary Triangulations. We now apply binary triangulations in a least squares approximation scheme following the principles of the general Algorithm 8.1. We base the algorithm on two important observations from the discussion of binary triangulations in Section 2.10.

i) the triangle vertices at any level k of a binary triangulation constitute a subset of a regular grid Ψ_k, and

ii) the grid Ψ_{k+1} at the next finer level is obtained by inserting grid lines halfway between the grid lines of Ψ_k.

Let Ψ_1 denote the regular grid defined at the initial level depicted on the left of Figure 8.13. There are 2×2 rectangular grid cells, or 3×3 vertices in Ψ_1, but only five vertices (black bullets) are active in the initial triangulation Δ_1. We proceed to the next level as follows: for each grid cell in Ψ_1 a criterion for subdivision is examined, for example by using an error measure or a counting measure as explained above. If subdivision is required within the grid cell, the vertex in the middle of the cell belonging to the next finer grid Ψ_2 is activated. Consequently, its two parent vertices in Ψ_1 must also be activated to avoid cracks in the triangulation Δ_2. This is illustrated on the left in Figure 8.13 by the insertion of vertex $v_{3,1}^2$ of Ψ_2 which causes insertion of the two parents, $v_{1,0}^1$ and $v_{2,1}^1$. The grid Ψ_2 at the next finer level has 4×4 grid cells obtained by inserting grid lines halfway between the grid lines in Ψ_1. When activating vertices in Ψ_3 to produce Δ_3 on the right, we observe that vertices belonging to both Ψ_2 and Ψ_1 must be activated to obtain a valid triangulation in this particular example.

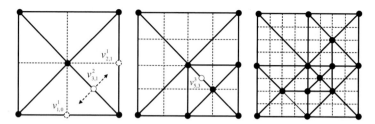

Fig. 8.13. Starting with the triangulation Δ_1 and the regular grid Ψ_1 on the left, vertex $v_{3,1}^2$ in Ψ_2 is first activated together with its two parents $v_{1,0}^1$ and $v_{2,1}^1$ in Ψ_1. The resulting triangulation is Δ_2 in the middle. Next, $v_{5,3}^3$ in Ψ_3 is activated together with ancestors belonging to both Ψ_1 and Ψ_2 to obtain Δ_3 on the right.

In general, at each level k of the subdivision scheme, the triangulation Δ_k is defined by a subset of the vertices of Ψ_k, which has $2^k + 1$ vertices in each direction,

$$\Psi_k = \{v_{i,j}^k\}_{i,j=0,0}^{2^k,2^k}.$$

When activating a vertex $v_{i,j}^k$ of Ψ_k, its parents in Ψ_{k-1} and all ancestors belonging to coarser grids $\Psi_{k-2}, \ldots, \Psi_1$ will also be activated if they are not already active. If $v_{i,j}^k$ is a vertex in the triangulation, then $v_{i,j}^k$ is active, and so are all ancestors of $v_{i,j}^k$ belonging to coarser grids. This is a necessary and sufficient condition for the triangulation to be valid without cracks.

At each level k of the algorithm, the system of equations in (8.18) can be solved by the Gauss-Seidel method or another iterative equation solver, for example the Jacobi method or the Conjugate gradient method. These equation solvers are easy to implement. The main effort lies in initializing the system matrix and the right-hand side from the input data overlaid the current triangulation. In particular, care must be taken when including the smoothing terms, which are required here to guarantee a unique solution.

Since an iterative equation solver is used, *nested iteration* can be applied where the resulting surface $f_k \in S_1^0(\Delta_k)$ at one level is used to supply an initial guess for $f_{k+1} \in S_1^0(\Delta_{k+1})$ at the next level. Since all vertices of the triangulation Δ_k are also in Δ_{k+1}, all coefficients from f_k can be transferred directly to the next level as starting values for the unknown coefficients of f_{k+1}. Moreover, each coefficient c_{ij}^{k+1} of f_{k+1} corresponding to a vertex (x_i, y_j) in Δ_{k+1} that is not in Δ_k, can simply be initialized by evaluating $f_k(x, y)$ which has been computed at level k,

$$c_{ij}^{k+1} = f_k(x_i, y_j).$$

So, when starting the iterative equation solver at the next finer level, there is already a good initial guess for the unknown coefficients obtained from the coarser level. Therefore, one may expect that relatively few iterations by the iterative equation solver are necessary to obtain a sufficiently accurate result.

We may note at this point that since the triangulations generated by the binary subdivision scheme are nested, the function spaces defined over them constitute a nested sequence of subspaces,

$$S_1^0(\Delta_1) \subset S_1^0(\Delta_2) \subset \cdots \subset S_1^0(\Delta_h),$$

where Δ_h is the triangulation at the finest level.

In the numerical examples presented in the remainder of this chapter, these general guidelines have been followed: at the first levels of the multilevel scheme an exact solution of the equation system is found by a direct solver. This ensures a good "take-off" by establishing a good global trend of the first surface as a basis for successive improvements when iterating at the finer levels with more unknowns. A combined stopping criterion based on relative improvement of the solution from one iteration to the next, and decrease of the residual $\left\| \mathbf{B}^T \mathbf{z} - \left(\mathbf{B}^T \mathbf{B} + \lambda \mathbf{E} \right) \mathbf{c} \right\|_2$ of Equation (8.18), has been used for the iterative solver. In most cases, between 10 and 20 iterations at each level is sufficient. Any further iterations do not improve the solution significantly when the default smoothing parameter λ_d in (8.19) is used. But with a larger smoothing parameter the number of iterations at each level is typically higher, e.g. up to 200 when λ is between 500 and 1000 times the default λ_d. These large values of λ were necessary in some of the examples to obtain a sufficiently smooth surface when the input data contained noise. The Gauss-Seidel method performed better than the Conjugate gradient method in this setting, even though the Conjugate gradient method converges faster when applied on a fixed mesh without a good initial guess.

A natural improvement of this simple coarse-to-fine ascending scheme is to use a *multigrid* scheme for solving the equation system [13]. Implementing a multigrid solver might be a difficult task, but by using so-called algebraic multigrid, one can download generic software from the Internet, for example a library called ML [44]. More details on the performance of using the ML library for solving our approximation problem can be found in [43]. For details on the theory of algebraic multigrid, see for example [12] and [11].

8.11 Numerical Examples for Binary Triangulations

The numerical examples below are based on four different data sources: data sampled from Franke's test function given by (8.14), real data from a terrain consisting of a combination of hypsographic data (contour data) and scattered measurements from the terrain surface, real scattered data from a mountain area in Norway, and lastly a data set with 3D parametrized data. The smoothing term is based on the thin-plate energy in all examples.

Approximation of data sampled from Franke's function. The scattered data set from Franke's test function consists of 3000 points sampled over the unit square. The points are unevenly distributed in the domain with relatively more data in areas with steep gradient or high curvature, see Figures 8.14 and 8.15. A combined subdivision criterion based on a counting measure and an error measure is used. A grid cell is thus refined if more than two points are inside the grid cell and the error for at least one of the points inside the cell is greater than the prescribed tolerance. The tolerance was 0.25 percent of $|z_{max} - z_{min}|$ of the given data, and the default value for λ was used. The algorithm terminated after subdivision of the grid Ψ_7 and delivered a triangulation Δ_8 with 1476 vertices and 2866 triangles. The number of Gauss-Seidel iterations at each level was between 9 and 12 to reach the stopping criterion. As expected, there are more triangles in areas with large curvature and high density of data due to the combined error and counting measure used as subdivision criterion. Also note the nice spatial grading from small triangles to larger triangles in Figure 8.15. If the middle vertex of each grid cell were activated when operating on Ψ_7, the resulting triangulation would have 33025 vertices, which also would be the number of unknowns in the equation system at that level. Thus, less than 4.5 percent of the maximum number of available vertices are used in the triangulation.

Figure 8.16 demonstrates the effect of choosing a very large λ (10 000 times λ_d), and thus demanding much smoothing. The surface leaves the given

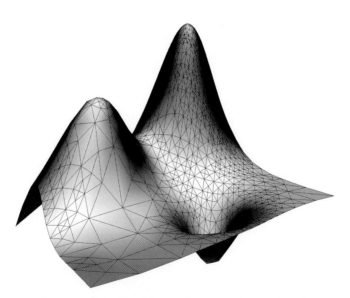

Fig. 8.14. Approximation to Franke's test function; the resulting binary triangulation imposed on the surface.

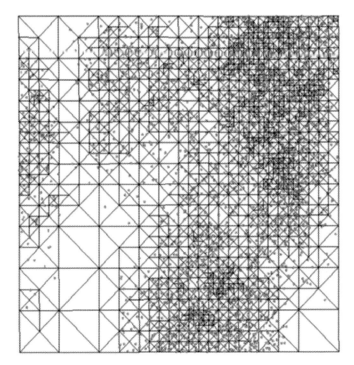

Fig. 8.15. The same binary triangulation as in Figure 8.14, and input data for numerical examples. The lower left corner corresponds to the nearest corner in Figure 8.14.

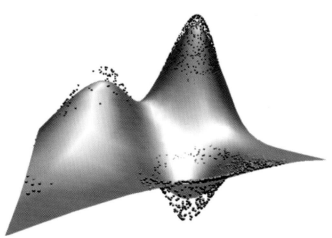

Fig. 8.16. Approximation with huge smoothing parameter.

data points, and for even larger λ when the smoothing term becomes more dominant, the surface approaches a plane. Also recall that the mean square error given by (8.28) increases monotonically towards a maximum with increasing λ.

Approximation of noisy data from Franke's function. Normally distributed noise was added to the data set used in the previous example. Subdivision of a grid cell was performed when there were more than two points inside the cell, but no error measure was used. To obtain a smooth pleasant-looking surface comparable to the surface produced in the previous example, λ had to be increased to 600 times λ_d. The number of Gauss-Seidel iterations at each level was between 73 and 207. The algorithm terminated at the same level as in the previous example. Figure 8.17 shows the approximation and the given noisy data points.

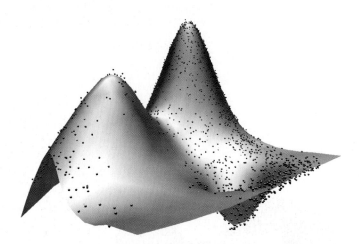

Fig. 8.17. Approximation to Franke's function from a data set with noise.

Terrain modeling from hypsographic data and scattered data. The terrain model shown in Figures 8.18 and 8.19 was derived from approximately 50 000 points consisting of both hypsographic data and scattered data points measured from the underlying terrain. An error measure was used as a subdivision criterion although the data contained noise. An acceptable smooth surface was obtained with a smoothing factor 10 times λ_d. The number of Gauss-Seidel iterations at each level was between 41 and 593. With an error tolerance of 2.5 percent of $|z_{\max} - z_{\min}|$, the algorithm terminated after subdivision at level 11 and delivered the triangulation Δ_{12} with 31204 vertices. The triangulation shown in Figure 8.19 was produced using a larger tolerance to

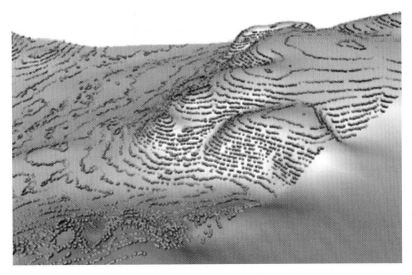

Fig. 8.18. Approximation of terrain data and the given hypsographic data.

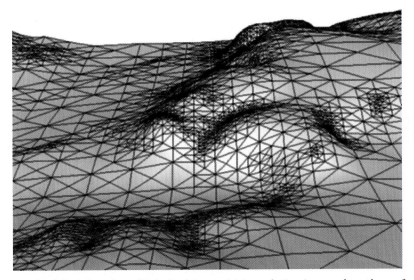

Fig. 8.19. Approximation of terrain data, and triangulation imposed on the surface.

avoid too many triangles in the presentation. The mesh is finer in areas with rapidly varying topography, and thereby captures the necessary details. We also observe the natural extrapolation of the surface to areas without input data, which is due to the thin-plate energy. Algorithms with good extrapolation properties are important in many applications. For example, in geological

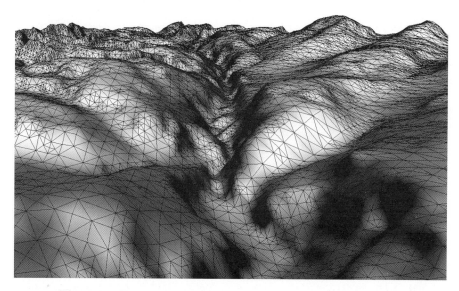

Fig. 8.20. Terrain modeling of an area in Jotunheimen, Norway.

modeling, faults and horizons must be extended to intersect each other with clean cuts outside their initial domain when creating boundary-based volume models [74].

Terrain modeling of mountain area. Figure 8.20 demonstrates another example of terrain modeling from a mountainous area in Jotunheimen in the west of Norway. The main point to notice here is that the binary triangulation through successive refinements by the approximation scheme automatically captures mountain peaks, ridges and valleys with triangles of appropriate size as needed, while simultaneously approximating the given scattered data by penalized least squares.

Approximation of parametrized 3D scattered data. A useful application of the adaptive properties of the multilevel scheme is demonstrated in Figure 8.21. The 2D data points shown in the parameter space on the left are parametrizations of 3D scattered data points sampled from the object on the right. Thus, a unique one-to-one mapping exists between the sampled data points in 3D space and the points in the 2D parametric space. This combined representation in 2D and 3D can be thought of as a set of quintuples $\{(u_i, v_i, x_i, y_i, z_i)\}_{k=1}^m$. The mapping, that is, the (u_i, v_i)-values in the parameter space were computed by a method called "shape-preserving" parametrization by Floater [29]. A characteristic of this method is that nearby points in 3D space are mapped closer and closer together in the parameter space as the distance between the 3D scattered data and the boundary of

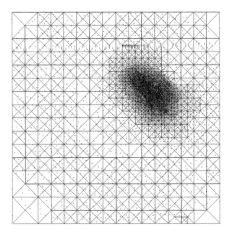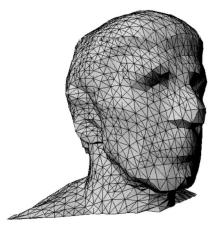

Fig. 8.21. Approximation of parametrized 3D scattered data.

the 3D object increases. The boundary of the 3D object is the opening of the sculpture in the bottom. This boundary was mapped to the boundary of the rectangular parameter domain. The samples in the example were relatively uniformly distributed on the 3D object, while the parametrized points end up in a cluster in the parameter space due to the characteristic of the mapping method.

Each of the space dimensions x, y and z was treated separately at each level k of the approximation scheme. That is, the quintuples were regarded as three different data sets $\{(u_i, v_i, x_i)\}_{i=1}^m$, $\{(u_i, v_i, y_i)\}_{k=1}^m$ and $\{(u_i, v_i, z_i)\}_{i=1}^m$, and solution vectors $\mathbf{c}_x$, $\mathbf{c}_y$ and $\mathbf{c}_z$ were found for the three space dimensions x, y and z respectively. These solution vectors represent the x, y and z-values respectively of the vertices of a 3D triangulation approximating the original samples on the 3D object. A counting measure was used as a subdivision criterion (based on number of points in the parameter space) to ensure that the same binary triangulation was used for each space dimension at each level. The binary triangulation at the finest level obtained from the subdivision scheme is shown on the left in Figure 8.21, and on the right the mapping of the binary triangulation to 3D space is shown as a 3D triangulation. While the sizes of the triangles in 2D parameter space vary considerably, we observe that they have approximately the same size when mapped back to 3D space.

8.12 Exercises

1. Show that the normal vector to a triangle patch can be expressed as in (8.5).

2. Let $m = n$ in the approximation problem of Section 8.2. Show that (8.13) reduces to $\mathbf{Bc} = \mathbf{z}$.
3. In Equation (8.16), verify that:
   a) all diagonal elements are non-zero,
   b) if $\Omega_i \cap \Omega_j \neq \phi$, then $\left(\mathbf{B}^T\mathbf{B}\right)_{ij}$ is zero if no points from P fall strictly inside $\Omega_i \cap \Omega_j$; and
   c) all non-zero elements of the system matrix $\left(\mathbf{B}^T\mathbf{B}\right)$ are positive.
4. Let $\mathbf{B}$ be an $m \times n$ matrix.
   a) Show that $\mathbf{B}^T\mathbf{B}$ is symmetric and positive semidefinite.
   b) Show that $\mathbf{B}^T\mathbf{B}$ is positive definite if $\mathbf{B}$ has linearly independent columns.
5. Show that the differentiating of $I(\mathbf{c})$ in Section 8.5 leads to the normal equations (8.18).
6. Show that the system matrix $\mathbf{B}^T\mathbf{B} + \lambda\mathbf{E}$ in (8.18) is positive definite when $\mathbf{B}^T\mathbf{B}$ is positive definite, $\mathbf{E}$ is positive semidefinite and $\lambda \geq 0$.
7. Give the details for arriving at the coefficients β_s^k, β_t^k, β_l^k, and β_r^k of the discrete thin-plate energy $\mathcal{T}_k f$ in the formula (8.27).
8. Derive the normal equations for "constrained weighted least squares". That is, combine weighted least squares in Section 8.8 with constrained least squares in Section 8.9 and find the matrix elements $\left(\mathbf{B}^T\mathbf{B}\right)_{ij}$ and the elements of the right-hand side $\left(\mathbf{B}^T\mathbf{z}\right)_i$ of the equation system.

9

Programming Triangulations: The Triangulation Template Library (TTL)

Triangulations can be dealt with algebraically using the principles of G-maps introduced in Chapter 2. High-level abstraction of functions operating on triangulations is achieved using G-maps, which are algebraically defined based on a limited number of clear concepts. At an abstract level, the topology of a triangulation can be described by using only one single topological element, the *dart*. Furthermore, the three α-iterators, α_0, α_1 and α_2, are the only operations needed for traversing the triangulation.

In this chapter, we utilize this simple concept to develop a *generic* triangulation library that we call TTL, "Triangulation Template Library". The meaning of the word "generic" is well known for programmers who are familiar with the Standard Template Library (STL), which is part of the C++ programming language. For others, this will become more clear in the following, although we do not go into details on formal design processes of generic programming in this text. The interested reader should consult one of the many text books in this field, for example [7]. However, we may already anticipate that algorithms in TTL are generic in the sense that they can operate on arbitrary data structures for triangulations through an adaptation layer which serves as an interface to the actual data structure used by the application. Thus, any data structure for triangulations, for example one of those from Chapter 2, can be used in combination with TTL if an appropriate adaptation layer is provided by the application programmer.

TTL with source code, documentation, demonstration programs and detailed programming instructions can be downloaded from the Internet, see [5]. An example where TTL is used to generate triangulations for efficient visualization of terrain is the soaring flight simulator *Silent Wings* [3], see front cover of this book.

9.1 Implementation of the Half-Edge Data Structure

When using object-oriented programming languages such as C++ and Java for implementing data structures for triangulations, it is natural to think of the topological entities vertices, edges and triangles as *classes*. For example, if we decide to define all topological entities as classes, the half-edge data structure introduced in Section 2.7 would need a node class, a half-edge class and possibly a triangle class. Figure 9.1 shows the pointer structure for the half-edge data structure, and Figure 9.2 depicts a possible class diagram for the half-edge data structure where a triangulation class has also been added.

The triangle class has a reference to one of its half-edges, a half-edge class has a reference to its source node and references to the next half-edge in the same triangle (in counterclockwise order), and to its twin-edge in the

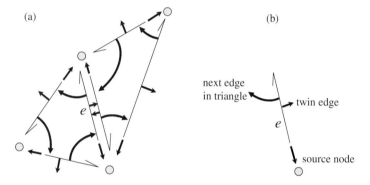

Fig. 9.1. The pointer structure of two triangles of a triangulation for the half-edge data structure. In (b), the three references from the half-edge e in (a) are shown.

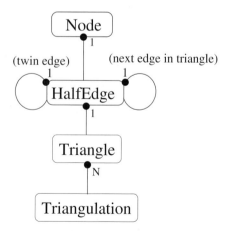

Fig. 9.2. Class diagram for the half-edge data structure.

adjacent triangle. The node class has no references to edges or triangles, but may carry its own geometric positional information in the Euclidean space. A triangulation class is also present and refers to a list with all triangles in the triangulation.

Let us exemplify implementation of the half-edge data structure in the C++ programming language from this simple object-oriented design. Only part of the code is shown here as we want to limit the space used for code listings. The reader should also note that the code in a professional implementation in a commercial product would look quite different. Also, the object-oriented design may differ from the simple design in Figure 9.2. Here we just want to show the principles of a possible implementation sufficient for later use when adapting the data structure to the generic functions in TTL. Only basic features of C++ that are well known to programmers are used.

The data members of the node class will naturally be the coordinates of the node in 3D space. These data are typically hidden in the private section of the class, so we need access functions in the public section of the class to retrieve the coordinates. The class, with implementation of its member functions inline, may read as follows:

```
class Node {
private:
    double x_, y_, z_;
public:
    Node(double x, double y, double z) {x_=x;y_=y;z_=z;}
    double& x() {return x_;}
    double& y() {return y_;}
    double& z() {return z_;}
    ...
};
```

The edge class may store references to the other elements as described above in its private section. Member functions getSourceNode, getNextEdge InTriangle, and getTwinEdge in the public section of the class may be called to access data members from outside the class. These functions return copies of the C++ pointers which are used as references in the private section.

A half-edge has its counterpart in a G-map as a dart, say d. Thus, the function GetNextEdgeInTriangle has its counterpart in the G-map as the composition $\alpha_1 \circ \alpha_0(d)$, and likewise, the function getTwinEdge has its counterpart in the composition $\alpha_0 \circ \alpha_2(d)$.

In the implementation shown here we do not use a triangle class. A triangle is just referred to as one of its three half-edges. The notion of *leading edge* is used for this purpose indicating which of the three half-edges of a triangle that represents the triangle. The function isLeadingEdge() of the half-edge class

```
class HalfEdge {
private:
  Node* sourceNode_;
  HalfEdge* twinEdge_;
  HalfEdge* nextEdgeInTriangle_;
  ...
public:
  HalfEdge() {...}
  ...

  HalfEdge* getTwinEdge() {return twinEdge_;};
  HalfEdge* getNextEdgeInTriangle() {return nextEdgeInTriangle_;}
  Node* getSourceNode() {return sourceNode_;}

  bool isLeadingEdge() {return isLeadingEdge_;}
  ...
};
```

returns **true** for one of the three half-edges of a triangle and **false** for the other two. Thus, the collection of all triangles in a triangulation is represented as a sequence of leading edges. We represent the sequence by a **std::list** type in C++. The prefix **std::** stands for the *name space* **std**, which is the scope where "standard" types and algorithms in C++ are defined.

```
class Triangulation {
protected:
  std::list<HalfEdge*> leadingEdges_;
  ...

public:
  Triangulation() {}
  void createDelaunay(...);

  HalfEdge* initTwoEnclosingTriangles(...);

  ...

  void swapEdge(...);
  HalfEdge* splitTriangle(...);
  ...
  ...
};
```

We may also have various utilities in the classes such as functions for printing their contents and functions to support visualization of the triangulation.

Only the amount of data (typically in the private section of the class) and not the number of functions in a class, affects its use of memory.

An important task that is not detailed here is reference counting of objects that are shared between other objects. For example, a node object is shared between all half-edges that have that node as a source node. In C++ we need a mechanism (often referred to as *smart pointers* and *handles*) that helps us to get rid of objects that are no longer used. In Java, however, this is handled automatically by Java's garbage collector.

Object-oriented design and implementation of data structures for triangulations must be done with great care. It might well turn out that simple linear array structures, commonly used in plain C or Fortran, are more efficient regarding CPU-time than an object-oriented approach. In particular, one should avoid creating millions of small objects (nodes and half-edges) scattered around in the memory of the computer. This may slow down access operations since objects must be retrieved from arbitrary locations in the computer's memory. Creating and deleting these objects may also be time consuming. There are workarounds, but what works well with one compiler and hardware platform may perform worse and behave totally differently on another platform. Thorough analysis of design and extensive testing of the software during implementation are necessary to avoid these pitfalls.

9.2 The Overall Design and the Adaptation Layer

Triangulation software is customarily implemented in a static manner in the sense that algorithms are adapted to one specific data structure only. The consequences are that library functions must be totally reimplemented if the underlying data structure is changed, or if one wants to operate on different data structures in the same application. The reason for this rigid approach has probably been the lack of an abstract description of the topology structure and of functions operating on the topology.

G-maps provide a formal algebraic approach to model general boundary-based topological models (B-reps), for example triangulations. Its algebraic definition is generic in the sense that it does does not depend on any underlying data structure. This suggests that algorithms operating on triangulations can be based on the algebraic concepts of G-maps and implemented generically, clearly separated from the data structure that is actually used to represent the triangulation.

This chapter follows up the introduction to G-maps in Chapter 2 and utilizes G-maps as a robust and vigorous algebraic tool to model the topology of triangulations. The concept of *function templates* in the C++ programming language is employed to implement the generic TTL library. TTL comprises a set of queries and modifiers acting on the topology of the triangulation. Iterator concepts based on G-maps' α-iterators support traversal of the topological

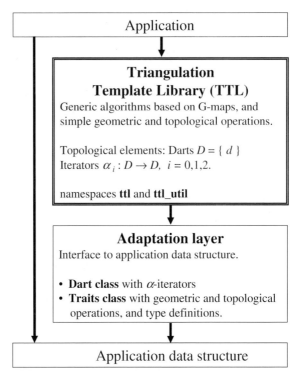

Fig. 9.3. An application using generic algorithms in TTL that act on the application's data structure through an adaptation layer.

structure similar to how basic pointer arithmetic is used in STL for traversal of linear sequences. Delaunay triangulation, including insertion of constrained edges to support constrained Delaunay triangulation, is implemented generically in TTL.

The simple principle of TTL and its interplay with an application is depicted in Figure 9.3. The arrows indicate information flow between the application, with its specific data structure, and TTL. TTL communicates with the application data structure through an adaptation layer, which is provided by the application programmer. The adaptation layer must implement a set of requirements required by function templates in TTL that are used by the application. Two components are requested in the adaptation layer:

- a *Dart class* (or struct) with α-iterators as described in Section 2.2, and
- a *traits class* (or struct) with basic geometric calculations, topological modifiers and type definitions.

Both components act on the application data structure. We explain this in detail in the next sections by using the half-edge data structure to exemplify.

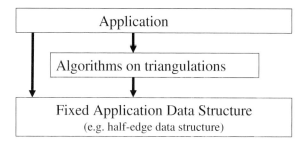

Fig. 9.4. Rigid non-generic design of triangulation software.

TTL has a name space called `ttl` where all functions and classes of TTL belong. A function in the `namespace ttl`, say `locateTriangle`, is then called by the syntax,

$$ttl::locateTriangle(<parameters>).$$

There are also some other name spaces in TTL available for the application programmer. For example `ttl_util` contains a set of utilities such as functions for creating random points, and basic point and vector algebra such as scalar product and cross product between vectors.

The main point to notice here is that the TTL library is totally independent of the underlying data structure. This is in contrast to a rigid non-generic design of triangulation software where algorithms work directly on a specific data structure as indicated in Figure 9.4.

9.3 Topological Queries and the Dart Class

Topological queries consist of functions that act on the topology without changing the topology or the geometry of the triangulation. A number of such functions are needed in commercial triangulation software, such as functions for accessing edges incident with a given node, functions for finding the triangles adjacent to a given triangle, functions for traversing the boundary of the triangulation, etc. These queries can easily be implemented using the concepts of darts and α-iterators in G-maps.

We start with a very simple function, that of deciding if an edge of a triangle is at the boundary of the triangulation. Recall from the algebraic description of G-maps that we allowed for fixed points for the α_2-iterator. Thus, if an edge e_j of a dart $d = (v_i, e_j, t_k)$ is at the boundary, then $\alpha_2(d) = d$. A pseudo-code for a function `isBoundaryEdge` that determines if e_j is a boundary edge can then be written simply as:

$$if \ \alpha_2(d) \ == \ d$$
$$\qquad return \ TRUE$$
$$else$$
$$\qquad return \ FALSE$$

Note that the behavior of this function and its return value is not affected by how the incoming dart is implemented, that is, how nodes, edges and triangles of the triangulation are represented in the actual data structure. This suggests that a C++ function template defines isBoundaryEdge [60]. If, for a certain data structure, the dart is defined as a class in C++ and the α_2-iterator is implemented as a member function alpha2 of the dart class, then the algorithm above can be implemented generically using a function template parametrized by dart type:

```
namespace ttl {

  template <class DartType>
  bool isBoundaryEdge(const DartType& d) {
    DartType d_iter = d;
    return (d_iter.alpha2() == d);
  }
};
```

This generic function can be used by applications based on any data structure for triangulations if a proper implementation of a dart class is provided by the application as an interface to the actual data structure. The function template expects that a member function alpha2 is present in the dart class and should return a dart as defined by the α_2-iterator. In addition, the function template expects that the dart class has a copy constructor and a boolean operator== for comparing objects. According to the definition of a fixed point, alpha2 should leave the dart unchanged if the edge associated with the dart is at the boundary of the triangulation.

Similarly, a family of other topological queries can be implemented as function templates. Different queries will require different member functions in the dart class. But in general, for topological queries where the topology of the triangulation is not changed, only the α-iterators need to be present, in addition to standard class member functions such as constructors, assignment operators and the like. The code listing in Figure 9.5 shows an example of a dart class that can serve as an interface between TTL and the half-edge data structure in the adaptation layer (Figure 9.3). The member functions alpha0, alpha1 and alpha2 correspond to the iterators α_0, α_1 and α_2 respectively. It is assumed that a class HalfEdge is implemented with member functions getNextEdgeInTriangle and getTwinEdge as explained in Section 9.1.

```
class Dart {
private:
  HalfEdge^ edge_;
  bool  dir_;

public:

  // Constructors and destructors
  ...
  Dart& operator= (const Dart& d) {. . .} // assignment

  bool operator==(const Dart& d) const {
    if (d.edge_ == edge_ && d.dir_ == dir_)
      return true;
    return false;
  }

  bool operator!=(const Dart& d) const {
    return !(d == *this);
  }

  Dart& alpha0() {dir_ = !dir_; return *this;}

  Dart& alpha1() {
    if (dir_ == true) {
      edge_ = edge_->getNextEdgeInTriangle()->getNextEdgeInTriangle();
    else
      edge_ = edge_->getNextEdgeInTriangle();
    dir_ = !dir_;
    return *this;
  }

  Dart& alpha2() {
    // Check if the dart is on the boundary. If yes, the dart
    // will not be changed.
    if (edge_->getTwinEdge()) { // Check if on boundary
      edge_ = edge_->getTwindEdge();
      dir_ = !dir_;
    }
    return *this;
  }
};
```

Fig. 9.5. Example of a C++ dart class for the half-edge data structure.

The data members of the dart class consist of a pointer to a half-edge and a boolean variable indicating whether the dart is positioned at the source node of the half-edge or at the target node of the half-edge. The α-iterators change the content of the dart and then return a reference to the dart itself. The latter is for convenience, as the function templates become more compact and are easier to write, see **isBoundaryEdge** above. Thus, the dart is implemented as a dynamic element which alters its content and "changes" its position in the triangulation through the α-iterators. The member function `alpha0`, which corresponds to the α_0-iterator, just switches the boolean variable `dir_` indicating that the dart is repositioned to the other node of the edge. The member functions `alpha1` and `alpha2` need some more operations to carry out the α_1 and α_2 operations. Note that `alpha2` assumes that `HalfEdge::getTwinEdge` returns a `NULL` pointer if the edge is at the boundary of the triangulation. This corresponds to a fixed point of the α_2-iterator.

The simple exercise outlined by `isBoundaryEdge` above also applies for other boundary based models where faces, bounded by edges, are not necessarily triangles. The only assumption in the dart class about faces being triangles is the member function `alpha1`, which must be modified slightly to handle faces with an arbitrary number of edges.

In addition to `isBoundaryEdge`, a number of other topological queries are needed in an application. The following simple function templates in `namespace ttl` are parametrized by dart type only and can operate on any data structure for triangulations, provided that a proper dart class exists in the adaptation layer.

```
bool isBoundaryTriangle(const DartType& d);
```

The given dart $d = (v_i, e_j, t_k)$ is an arbitrary dart positioned in the triangle t_k to be examined. Edges are checked as in `isBoundaryEdge` above and the composition $\alpha_1 \circ \alpha_0(d)$, which corresponds to a 2-orbit (see Definition 2.2 in Section 2.2), is used to reposition the dart to the next edge in the triangle. The triangle is found to be on the boundary if one of the three half-edges is on the boundary.

```
bool isBoundaryNode(const DartType& d);
```

The composition $\alpha_1 \circ \alpha_2(d)$, which corresponds to a 0-orbit, repositions the dart to an edge of the next triangle that has the node of the given dart as a member. The node is at the boundary of the triangulation if one of the edges is at the boundary. Each edge is checked as in `isBoundaryTriangle` above.

```
int getDegreeOfNode(const DartType& d);
```

We remember that the degree (or valency) of a node is the number of edges incident with the node. The number of incident edges equals the number of $\alpha_2 \circ \alpha_1$ compositions that can be done in the 0-orbit until the given dart is reached again. This function is slightly more involved than the two former since the node may be at the boundary of the triangulation, in which case a fixed point for the α_2-iteration is reached. Assuming that the actual node is not at the boundary, a function `getDegreeOfInteriorNode` can be implemented as follows:

```
namespace ttl {

  template <class DartType>
  int getDegreeOfInteriorNode(const DartType& d) {
    DartType d_iter = d;
    int degree = 0;
    do {
      d_iter.alpha1().alpha2();
      ++degree;
    } while (d_iter != d);
    return degree;
  }

};
```

Other queries commonly present in triangulation software are functions that return topological elements from the triangulation structure. For example, three triangles t_1, t_2 and t_3 adjacent to a given triangle can be return by a function template as three darts d_{t1}, d_{t2} and d_{t3} that have t_1, t_2 and t_3 as members of their respective (node, edge, triangle)-triples. The declaration may read as follows:

```
void getAdjacentTriangles(const DartType& d,
DartType& dt1, DartType& dt2, DartType& dt3);
```

If the given triangle is at the boundary of the triangulation, there must be a mechanism for indicating that one or more of the returned darts represent "NULL objects". A default constructor may prepare such an object, and a boolean function `Dart::isNull()` may return a true value if a dart is a NULL object.

The class template `std::list` in C++ can be used if a sequence of topological elements should be given, or returned from a function template. Alternatively, the function template can be parametrized to allow for a more general sequence type than the specific `std::list`. Still, such function templates can operate with darts as the only topological type in their interfaces.

9.4 Some Iterator Classes

The k-orbits in Definition 2.2 represent circular sequences in the sense that when traversing the darts of an orbit in one direction the same dart is reached again (though this is not the case for 0-orbits and 1-orbits when a fixed point is reached at the boundary). Thus, they are different from linear sequences supported by container classes and iterators in the Standard Template Library (STL). A concept called *circulators* [47] can be used to operate with such circular sequences similar to how STL operates with iterators. Traversal of darts in a k-orbit can then be done using almost the same syntax as that used by STL when traversing standard sequences like `std::list` or `std::vector`. From an application programmer's point of view, many traversal operations are simplified when circulators are used. For example, an iterator for for advancing a dart in both directions of a 0-orbit can be defined as follows.

```
template <class DartType>
class Orbit0_iterator {
  DartType d_;
public:
  Orbit0_iterator(const DartType& d) : d_(d) {}
  ...
  Orbit0_iterator& operator++() {d_.alpha1().alpha2();
    return *this;}
  Orbit0_iterator& operator--() {d_.alpha2().alpha1();
    return *this;}
  DartType& operator*(){return d_;}
  const DartType& operator*() const {return d_;}
  ...
};
```

`Orbit0_iterator` is initialized through the constructor by a given dart d of the actual 0-orbit. The dart is advanced in the 0-orbit by `operator++` and `operator--` implemented by compositions $\alpha_2 \circ \alpha_1(d)$ and $\alpha_1 \circ \alpha_2(d)$, respectively. Thus, the iterator is *bidirectional*. Note that these operators advance the dart to every second dart position of the 0-orbit, unless the node of the 0-orbit is at the boundary of the triangulation. So, for a 0-orbit corresponding to a node in the interior of the triangulation, the dart has the same direction (clockwise or counterclockwise) inside each visited triangle. The `operator*` "dereferences" the iterator to the current dart of the iterator. Similarly, classes `Orbit1_iterator` and `Orbit2_iterator` for advancing a dart in a 1-orbit and a 2-orbit, respectively, would also simplify many algorithms in an application. (See also Exercise 6.)

Other examples of circular sequences are boundaries of holes and the exterior boundary of a triangulation. A class `Boundary_iterator` may implement an iterator for advancing a dart along the boundary of the triangulation.

```
template <class DartType>
class Boundary_iterator {
  DartType d_;
public:
  Boundary_iterator(const DartType& d) : d_(d) {}
  ...
  Boundary_iterator& operator++() {...; return *this;}
  Boundary_iterator& operator--() {...; return *this;}
  DartType& operator*(){return d_;}
  const DartType& operator*() const {return d_;}
  ...
};
```

Here, operator++ and operator-- may advance the dart from one boundary edge to another boundary edge while keeping the dart in the same direction inside each visited triangle as illustrated in Figure 9.6.

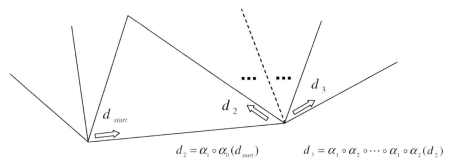

$$d_2 = \alpha_1 \circ \alpha_0(d_{start}) \qquad d_3 = \alpha_1 \circ \alpha_2 \circ \cdots \circ \alpha_1 \circ \alpha_2(d_2)$$

Fig. 9.6. Iteration at the boundary of a triangulation. The operator++ of the boundary iterator advances the dart d_{start} to d_3 (and operator-- takes dart d_3 "backwards" to d_{start}.

9.5 Geometric Queries and the Traits Class

In the previous sections, only topological information was used by the function templates, and topological operations required in the dart class were limited to α-iterators only. In this section, TTL is extended to use geometric embedding information of nodes in the triangulation structure. The geometric operations will still be at a query level in the sense that no changes of the triangulation take place through the operations. Topological and geometric operations that modify the triangulation will be discussed in Section 9.6.

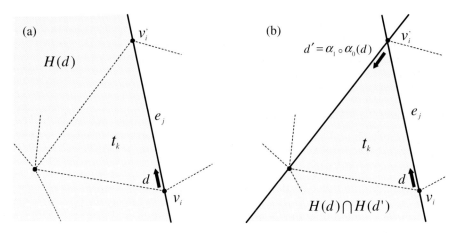

Fig. 9.7. Illustration of half-plane and intersection of half-planes. The patterned area in (a) is the half-plane $H(d)$ associated with the dart d. The patterned area contains v_i and v_i', and extends to infinity everywhere left of e_j. In (b), the intersection of the two half-planes associated with the darts d and $d' = \alpha_1 \circ \alpha_0(d)$ is shown.

A simple example involving geometric calculations is that of finding if a given point p is located inside a given triangle. Assume that a dart $d = (v_i, e_j, t_k)$ is oriented counterclockwise in the triangle t_k. Let $H(d)$ denote the half-plane to the left of d containing v_i and the opposite node v_i' in t_k, see Figure 9.7(a). Further, let d' denote the dart from the composition $\alpha_1 \circ \alpha_0(d)$. The intersection of the two half-planes $H(d) \cap H(d')$ defines an angular sector spanned by the two edges associated with d and d' as depicted in Figure 9.7(b), and the region defined by the intersection

$$H(d) \cap H\left(\alpha_1 \circ \alpha_0(d)\right) \cap H\left(\alpha_0 \circ \alpha_1(d)\right) \tag{9.1}$$

is exactly the triangle t_k. Thus, p is located in t_k if and only if p lies in the three half-planes of expression (9.1). The query $p \in H(d)$ can be implemented by evaluating the sign of a determinant. Let (x_1, y_1) and (x_2, y_2) be the positions in the plane of v_i and v_i', and let $p = (x_3, y_3)$. The determinant,

$$\begin{vmatrix} x_1 & y_1 & 1 \\ x_2 & y_2 & 1 \\ x_3 & y_3 & 1 \end{vmatrix} = (x_1 - x_3)(y_2 - y_3) - (x_2 - x_3)(y_1 - y_3),$$

evaluates to zero if p is on the line between v_i and v_i', it is greater than zero if p is to the left of the directed line from v_i to v_i', and less than zero otherwise. High level abstraction is obtained if such computational geometry functionality can be provided from the outside TTL. This suggests that the query $p \in H(d)$, say `bool inLeftHalfPlane`, should be implemented on the application side in an interface to the actual data structure. Along with the application's dart class, the function is naturally placed in the adaptation layer, see (Figure 9.3). One may encapsulate `inLeftHalfPlane`, and other basic geometric operations, in a *traits class*. The C++ standard has the following definition of a traits class:

Definition 9.1 (Traits class). *A traits class is a C++ class that encapsulates a set of types and functions necessary for template classes and template functions to manipulate objects of types for which they are instantiated.*

Thus, our traits class is an ordinary C++ class (or a C++ struct) where the application programmer implements functions required by function templates in TTL for operating on the actual data structure. It also contains type definitions, for example, it defines whether a real number should be represented as a `float` or a `double`. In addition, it contains topological operations that cannot be resolved by TTL, such as removing a triangle or swapping an edge. The traits class is passed as a template argument to those functions in TTL that require these functions. More on traits can be found in many text books or in [65].

So, by leaving the responsibility for geometric calculations to the application by directing them to the traits class, the application programmer can choose the level of accuracy for these operations. Tests like $p \in H(d)$ can be implemented simply as above by evaluating the sign of a determinant. But the test involves floating point arithmetic which may lead to incorrect results due to round-off errors when the determinant is near zero. The problem can be solved using exact arithmetic, but then the speed would probably be reduced by several orders of magnitude. Shewchuk [78] describes different techniques for solving the problem above and related problems involving point and vector algebra with different levels of accuracy. A function template `inTriangle`, which checks if a given point is located inside a given triangle, can be implemented in C++ thus:

```
namespace ttl {

    template <class TraitsType, class PointType, class DartType>
    bool inTriangle(const PointType& p, DartType& d) {
      for (int i = 0; i < 3; i++) {
        if (!TraitsType::inLeftHalfPlane(p, d))
          return false;
        d.alpha0().alpha1();
      }
      return true;
    }

};
```

The function template is parametrized by traits type, point type and dart type, and the triangle in question is given to the function as a dart that has the triangle as a member of its triple. The boolean function `inLeftHalfPlane`, declared and defined as required by `ttl::inTriangle`, must be present in the application's traits class. It takes a point p and a dart d as input, and p is tested to be in the half-plane defined by d as outlined above. It is assumed that d is oriented counterclockwise in the triangle. From the application side, `inTriangle` is called with the following syntax:

```
MyPoint p;
MyDart d;
...
bool in_triangle = ttl::inTriangle<MyTraits>(p,d);
...
```

where `MyTraits` is the name of the traits class for the actual data structure. If all members of the traits class (data members and member functions) are declared static, the class itself is static and need not be instantiated as an object. It can also be a struct as shown here:

```
struct MyTraits {
  ...
  static bool inLeftHalfPlane(const MyPoint& p, const
    MyDart& d);
  ...
};
```

The application programmer can now implement this function at the desired level of accuracy. In many cases it suffices just to calculate the sign of a determinant as explained above. Such standard functions with "default implementation" are also present in the TTL library in namespace `ttl_util` that can be called from the corresponding functions in the traits class,

```
namespace ttl_util {

bool inLeftHalfPlaneFast(const MyPoint& p, const MyDart& d);

};
```

Another function common in triangulation software is that of locating the triangle in a triangulation containing a given point. It can be implemented based on the same functionality in the adaptation layer (the dart class and traits class) as required by `inTriangle` above. Note first how fixed points for the α_2-iterator at the boundary of the triangulation can be handled. If for some dart d we have $\alpha_2(d) = d$ (a fixed point), then d is positioned at a boundary edge. If d is oriented counterclockwise in the triangle and the boundary of the triangulation is convex, then p lies outside the triangulation if p is not in the half plane of d,

$$\alpha_2(d) = d \text{ and } p \notin H(d) \iff p \text{ is outside the triangulation.} \qquad (9.2)$$

Algorithm 9.1 takes as input a point p in the plane and an arbitrary dart $d = (v_i, e_j, t_k)$ oriented counterclockwise in triangle t_k. It is assumed that a half-plane $H(d)$ is defined as above.

Algorithm 9.1 Dart `locateTriangle(Point` p`, Dart` d`, bool` *found*`)`

1. $d_{start} = d$
2. **if** $p \in H(d)$ // is p in the half-plane $H(d)$?
3. $d = \alpha_1 \circ \alpha_0(d)$ // next edge counterclockwise (ccw.)
4.   **if** $d == d_{start}$
5.   *found* = TRUE, **return** d // inside triangle of d
6. **else**   // try to move to the adjacent triangle
7.   **if** $\alpha_2(d) == d$ // check if on boundary
8.   *found* = FALSE, **return** d // outside triangulation
9. $d_{start} = \alpha_0 \circ \alpha_2(d)$
10. $d = \alpha_1 \circ \alpha_2(d)$ // next edge ccw. in adjacent triangle
11. **goto** 2

Figure 9.8 shows how the given dart d is successively repositioned until it reaches the triangle where p is located. The composition $\alpha_1 \circ \alpha_0(d)$ (Step 3) moves the dart inside a triangle, and with the composition $\alpha_1 \circ \alpha_2(d)$ (Step 10), the dart is moved across an edge of two adjacent triangles and positioned at the next edge. Since a composition of two α-iterators is used to move the dart, it is always kept in counterclockwise direction inside a triangle. The dotted darts in the figure represent d_{start} in Step 9. No triangle or half-plane associated with an edge is ever considered twice. The algorithm terminates with d in the located triangle, or it terminates at Step 8 with d at the boundary if p is outside the triangulation. In the latter case it is assumed that the boundary is convex and that there are no holes in the triangulation, in which case the relation (9.2) is employed. A triangle is also located if p is on an edge or on a node. The algorithm is fast, but it is not robust with respect to triangles that are degenerate or almost degenerate.

Many other similar algorithms have been described in the literature, see for example [20]. Some of them need a special data structure which is static in the sense that it is not longer valid if the topology of the triangulation is changed. But Algorithm 9.1 may operate on the actual data structure for the triangulation and implementation in the framework of TTL is generic. A robust algorithm based on a dynamic search structure that is updated while new points are inserted into the triangulation can be found in [35].

Implementation of `locateTriangle` in C++ is very similar to the pseudo-code:

```cpp
namespace ttl {

    template <class TraitsType, class PointType, class DartType>
    bool locateTriangle(const PointType& p, DartType& d_iter) {
      DartType d_start = d_iter;
      DartType d_prev;
      for (;;) {
        if (TraitsType::inLeftHalfPlane(p, d_iter)) {
          d_iter.alpha0().alpha1();
          if (d_iter == d_start)
            return true; // left to all edges in triangle
        }
        else {
          d_prev = d_iter;
          d_iter.alpha2();
          if (d_iter == d_prev)
            return false; // iteration to outside boundary

          d_start = d_iter;
          d_start.alpha0();
          d_iter.alpha1(); // counterclockwise in next triangle
        }
      }
    }

};
```

Similarly, many other methods commonly present in triangulation software can be implemented generically as function templates based on a limited number of primitive geometric operations required in the interface to the actual data structure. In addition to `inLeftHalfPlane`, scalar product and cross product between vectors[1] may suffice to cover a variety of geometric functions. Vectors can be represented by darts as will be demonstrated later.

---

[1] `inLeftHalfPlane` can also be regarded as a cross product calculation.

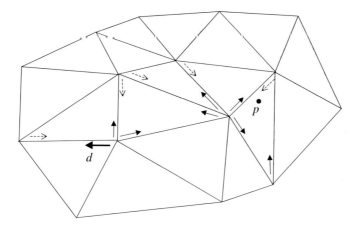

Fig. 9.8. Localizing a point p starting from a dart d.

9.6 Geometric and Topological Modifiers

So far the scope has been on generic function templates that are queries, in the sense that neither topology nor geometry are changed through the operations. Below we discuss how TTL can be extended to include operations that modify the topology and the geometric embedding information of triangulations.

In many applications, for example in terrain modeling, the only geometric embedding information of a triangulation is the 3D positions of nodes. Thus, a natural extension of the concepts above is to require elementary geometric operations in the adaptation layer, in the traits class or in the Dart class, that change positions of nodes. Together with corresponding access functions and topological query functions based on α-iterators, this would be sufficient for functions that modify the geometry of a triangulation.

Topological modifiers are more complex. They can roughly be divided into three main categories.

1. Modifiers that add nodes, edges and triangles in a triangulation.
2. Modifiers that remove nodes, edges and triangles. Removing one element may also imply removing some of the others. For example, removing an interior edge also implies removing two triangles.
3. Modifiers that preserve all nodes and the number of edges and triangles. This can be done by swapping edges that are diagonals of strictly convex quadrilaterals. As stated by Theorem 3.4, all possible triangulations of a set of points can be reached by a sequence of edge-swaps starting from an initial triangulation of the points. Inserting a constrained edge between two existing nodes is an example of this type of modifier, where the new constrained triangulation can be obtained by edge-swapping only (Section 6.5).

The two former modifiers that alter the number of nodes, edges and triangles are sometimes called Euler operations. This is because the number of nodes, edges and triangles is governed by the Euler Polyhedron formula, $|T| = |E| - |V| + 1$, that we developed in the preliminaries in Section 1.3.

Several extensions of TTL described previously are necessary to incorporate topological modifiers as generic functions, and more more effort will be necessary for the application programmer to implement counterparts in the adaptation layer for interfacing the actual data structure.

Topological modifiers can be described algebraically using topological operators based on the concept of *sewing* in G-maps. These operators establish the involutions in a G-map, that is, the relationships between darts as defined through α-iterators. Two darts d_i and d_j are said to be *k-sewed*, or α_k-*sewed*, in a G-map if $\alpha_k(d_i) = d_j$ (and $\alpha_k(d_j) = d_i$). We could suggest that topological modifiers should be implemented generically in TTL by using sewing as atomic operations, and require that sewing operations be implemented in the adaptation layer where they carry out the operations on the application data structure. Although elegant and conceptually in agreement with how α-iterators are handled, we would face problems with this approach.

Let us analyze edge-swapping algebraically through sewing operators. Figure 9.9 shows two triangles forming a convex quadrilateral in a triangulation. The edge e_i with nodes v_1 and v_2 can be swapped to become a new edge e'_i with nodes v_3 and v_4. Apart from assigning new nodes to e'_i, a total of six sewing operations are required to perform the edge-swapping: each unprimed dart d_i in the figure must be α_1-sewed with the primed dart d'_i for $i = 1, \ldots, 6$, to establish a correct topology. This involves all darts of the two triangles. Only α_1-sewings are involved in the swapping, α_0-sewings and α_2-sewings are all maintained.

An attempt to execute a function like $\texttt{sew1}(d_1, d'_1)$ separately on a data structure, that establishes the relationship $\alpha_1(d_1) = d'_1$ and $\alpha_1(d'_1) = d_1$, would destroy the pointer structure and create an intermediate topological representation that is not legal. We therefore leave the responsibility for the whole swapping operation to the application such that the six sewing operations can be carried out "simultaneously", together with other necessary updates of the data structure. Thus, topological modifiers in TTL that need swapping require a function $\texttt{swapEdge(DartType\& d)}$ in the traits class that swaps an edge associated with the given dart. The numbering of darts in Algorithm 9.2 refers to Figure 9.9.

Modifiers that remove nodes, edges and triangles from a triangulation are also handled by TTL, as well as modifiers that create and add topological elements. For example, for incremental Delaunay triangulation in the next section, the function $\texttt{ttl::insertNode(DartType\& d, NodeType\& n)}$ inserts node $\texttt{n}$ in the triangulation and then swaps edges until the triangulation is Delaunay. A function $\texttt{splitTriangle(DartType\& d, NodeType\& n)}$, which

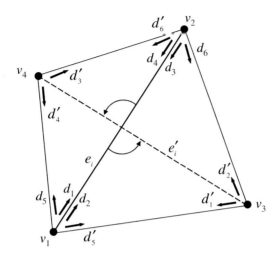

Fig. 9.9. Sewing-operations when swapping an edge. Each unprimed dart d_i must be α_1-sewed with the primed dart d_i' for $i = 1, \ldots, 6$.

Algorithm 9.2 swapEdge(Dart d_1)

1. $d_2 = \alpha_2(d_1)$, $d_5' = \alpha_1(d_2)$, $d_1' = \alpha_0(d_5')$, $d_2' = \alpha_1(d_1')$, $d_6 = \alpha_0(d_2')$, $d_3 = \alpha_1(d_6)$
2. $d_5 = \alpha_1(d_1)$, $d_4' = \alpha_0(d_5)$, $d_3' = \alpha_1(d_4')$, $d_6' = \alpha_0(d_3')$, $d_4 = \alpha_1(d_6')$
3. **for** $(i = 1, \ldots, 6)$
   sew1(d_i, d_i')

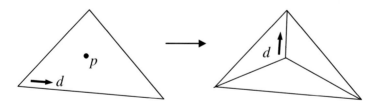

Fig. 9.10. Splitting a triangle at an insertion point into three new triangles.

inserts a node by splitting a given triangle associated with a dart into three new triangles as illustrated in Figure 9.10, is then reqired in the traits together with swapEdge(DartType& d) explained above.

9.7 Generic Delaunay Triangulation

In this section, we describe briefly how the incremental Delaunay triangulation algorithm of Chapter 4 is implemented generically in the framework of TTL. We remember that incremental Delaunay triangulation starts with an initial Delaunay triangulation and then inserts all points from a point set P one by one into the triangulation. After each point insertion, the triangulation is updated to be Delaunay, such that all triangles satisfy the circumcircle test. Here we let the initial triangulation be two (large) triangles enclosing all points of P (Figure 9.14(a)). Creating these triangles is the responsibility of the application; cf. the function initTwoEnclosingTriangles in class Triangulation of the half-edge data structure in Section 9.1. So basically, using TTL for incremental Delaunay triangulation, consists of repetitive calls to the function template ttl::insertNode that will be described below.

Recall from Section 4.6 the following steps of inserting a point p into an existing Delaunay triangulation Δ_N with N nodes to obtain a new Delaunay triangulation Δ_{N+1} with $N + 1$ nodes:

Step 1 Locate the triangle t in Δ_N that contains p.

Step 2 Split t into three triangles by making three new edges between p and the nodes of t, and thus obtain a new triangulation Δ'_{N+1}.

Step 3 Apply a swapping procedure based on the circumcircle test to swap edges that are not locally optimal in Δ'_{N+1} until all edges are locally optimal and the final triangulation Δ_{N+1} is Delaunay.

Figure 9.11(b) shows the situation after point p has been inserted into an existing Delaunay triangulation in Step 2 above, but before the swapping process in Step 3.

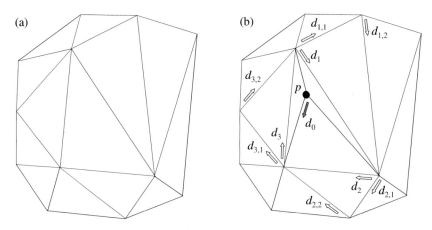

Fig. 9.11. Illustration for Algorithm 9.4 when starting the swapping procedure.

The pseudo-code that follows is based on the dart algebra of G-maps. Implementation in C++ is a straightforward exercise that can be done similarly to the examples in Section 9.3 and 9.5. It is assumed that a dart d, given as input or output from the generic algorithms in TTL, is always kept in counterclockwise direction inside a triangle. Note that some details, for example handling fixed points at the boundary, are omitted in the code listing. Algorithm 9.3 implements Step 1, 2 and 3 above. An insertion point p and an arbitrary counterclockwise dart d in the existing Delaunay triangulation Δ_N is given as input.

Algorithm 9.3 `insertNode(Point` p`, Dart` d`)`

---

1. $d_t = $ `ttl::locateTriangle`$(p, d, found)$
2. $d_0 = $ `app::splitTriangle`(d_t, p)
3. $d_1 = \alpha_2 \circ \alpha_1 \circ \alpha_0 \circ \alpha_1 \circ \alpha_2 \circ \alpha_1(d_0)$
4. $d_2 = \alpha_2 \circ \alpha_1 \circ \alpha_0 \circ \alpha_1(d_0)$
5. $d_3 = \alpha_2 \circ \alpha_1 \circ \alpha_2 \circ \alpha_0(d_0)$
6. `ttl::recSwapDelaunay`(d_1)
7. `ttl::recSwapDelaunay`(d_2)
8. `ttl::recSwapDelaunay`(d_3)

Again, the prefix `ttl::` at a function call indicates that the function is part of the generic TTL library, and the prefix `app::` indicates that the function must be present on the application side in a traits class as an interface to the actual data structure (Figure 9.3). In Step 1, Algorithm 9.1 in Section 9.5 is called to locate the triangle that contains the insertion point p. The located triangle is represented by a dart d_t that has the triangle as a member of its triple. Then the located triangle is split into three new triangles by `app::splitTriangle`, which also delivers a new dart d_0 oriented counterclockwise and located at the insertion point p (Figure 9.11). Next, the darts d_1, d_2 and d_3 shown in Figure 9.11 are found by compositions of α-iterators. Each dart is then passed to the recursive swapping procedure `recSwapDelaunay` in TTL, which was explained in detail in Section 4.6. In Algorithm 9.4 we use dart algebra and write the swapping procedure with composition of α-iterators. The numbering of darts is the same as in Figure 9.11.

In Step 1 the circumcircle test is applied to the current edge e_i associated with the dart d_i. As indicated by the prefix `ttl::`, it is assumed that the circumcircle test is part of TTL although its lower-level numerical calculations are directed to the application, see Algorithm 9.5 below. Recursion is stopped in Step 2 if e_i passes the circumcircle test, that is, if e_i is locally optimal. Otherwise, candidate edges for swapping in the next recursion, represented by the darts $d_{i,1}$ and $d_{i,2}$ opposite d_i, are found by compositions of α-iterators. After

---

Algorithm 9.4 recSwapDelaunay(Dart d_i)

---
1. if (ttl::circumcircleTest(d_i) == TRUE)
2.   **return**   // Do not swap the edge
3. $d_{i,1} = \alpha_2 \circ \alpha_1(d_i)$
4. $d_{i,2} = \alpha_2 \circ \alpha_0 \circ \alpha_1 \circ \alpha_0(d_i)$
5. app::swapEdge(d_i)
6. ttl::recSwapDelaunay($d_{i,1}$)   // call this procedure recursively
7. ttl::recSwapDelaunay($d_{i,2}$)   // call this procedure recursively

---

e_i is swapped in Step 5, the same procedure is called again recursively with $d_{i,1}$ and $d_{i,2}$. On termination of the algorithm, all edges of the new triangulation Δ_{N+1} are locally optimal, and it follows from Theorem 3.3 that Δ_{N+1} is a Delaunay triangulation. Figure 9.13 shows the whole swapping process of Algorithm 9.3 when inserting p. From (b) to the final triangulation in (e), each picture shows the triangulation after one new edge has been swapped. The notation of the darts are the same as in algorithm recSwapDelaunay.

Lastly, we outline implementation of the circumcircle test in TTL. We recall from Section 3.7 that an edge e_i which is a diagonal in a quadrilateral should be swapped if

$$\sin \alpha \cos \beta + \cos \alpha \sin \beta < 0,$$

where α and β are the angles opposite e_i in the quadrilateral as shown in Figure 9.12. To maintain a simple syntax for the generic interface to TTL, we may interpret the darts $\mathbf{d}_i$, $i = 1, 2, 3, 4$ in the figure as unit vectors, and as in Section 3.7 the vectors are defined in 3D space (but exist in 2D space with zero z-component). Then sine and cosine of α and β can be computed by cross products and scalar products,

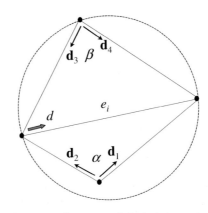

Fig. 9.12. Circumcircle test and darts interpreted as vectors.

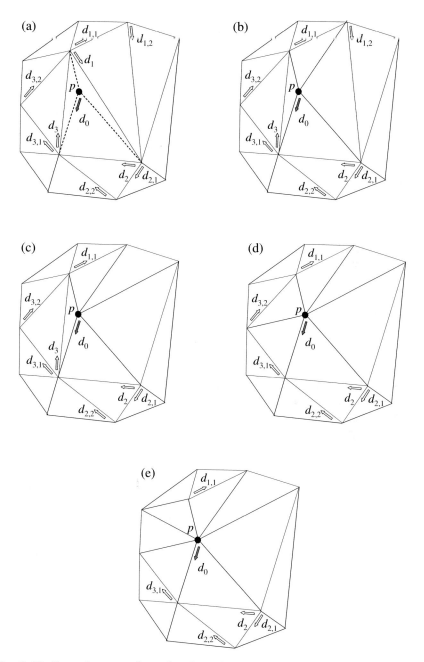

Fig. 9.13. Swapping procedure when inserting a point p into a Delaunay triangulation. From (b) to the final triangulation in (e), each picture shows the triangulation after one new edge has been swapped.

$$\sin \alpha = (\mathbf{d}_1 \times \mathbf{d}_2) \cdot \mathbf{e}_3$$
$$\sin \beta = (\mathbf{d}_3 \times \mathbf{d}_4) \cdot \mathbf{e}_3$$
$$\cos \alpha = \mathbf{d}_1 \cdot \mathbf{d}_2$$
$$\cos \beta = \mathbf{d}_3 \cdot \mathbf{d}_4,$$

where $\mathbf{e}_3 = (0,0,1)$. Assume that cross product and scalar product between vectors, represented as darts, are present on the application side. Let a function $\texttt{crossProduct}(\mathbf{d}_i, \mathbf{d}_j)$ deliver the scalar $(\mathbf{d}_i \times \mathbf{d}_j) \cdot \mathbf{e}_3$, that is, the sine of the angle between $\mathbf{d}_i$ and $\mathbf{d}_j$ in accordance with the equations above. Similarly, $\texttt{scalarProduct}(\mathbf{d}_i, \mathbf{d}_j)$ delivers the scalar $(\mathbf{d}_1 \cdot \mathbf{d}_2)$, which is the cosine of the angle between $\mathbf{d}_i$ and $\mathbf{d}_j$. The circumcircle test in Algorithm 9.5 takes a dart d representing a diagonal edge in a quadrilateral as input.

Algorithm 9.5 `bool circumcircleTest(Dart d)`

1. $\mathbf{d}_1 = \alpha_1 \circ \alpha_0 \circ \alpha_1 \circ \alpha_2(d)$
2. $\mathbf{d}_2 = \alpha_0 \circ \alpha_1 \circ \alpha_2(d)$
3. $\mathbf{d}_3 = \alpha_0 \circ \alpha_1(d)$
4. $\mathbf{d}_4 = \alpha_1 \circ \alpha_0 \circ \alpha_1(d)$
5. $\sin \alpha$ =`app::crossProduct`$(\mathbf{d}_1, \mathbf{d}_2)$
6. $\sin \beta$ =`app::crossProduct`$(\mathbf{d}_3, \mathbf{d}_4)$
7. $\cos \alpha$ =`app::scalarProduct`$(\mathbf{d}_1, \mathbf{d}_2)$
8. $\cos \beta$ =`app::scalarProduct`$(\mathbf{d}_3, \mathbf{d}_4)$
9. **if** $(\sin \alpha \cos \beta + \cos \alpha \sin \beta) < 0$
10.   **return** FALSE   // swap the edge
11. **else**
12.   **return** TRUE

In the actual implementation in TTL, Step 9 is replaced with the more robust test in Algorithm 3.1. There is also a mechanism in TTL to avoid cycling and infinite loops in nearly neutral cases. Due to these circumstances and other numerical instability that can occur, one could consider directing the whole circumcircle test and other functionality involving numerical calculations to the application (to the traits class in the adaptation layer). Another solution would be to always direct such functionality to the application, but also implement it in TTL (in namespace `ttl_util`). Then the application programmer would choose whether to implement the functions at the desired level of accuracy or call the functions already present in `ttl_util`. These mechanisms are implemented in TTL for `inLeftHalfPlane` used by `ttl::locateTriangle` described previously, and for lower-level point and vector algebra operations, such as scalar product and cross product between vectors.

Removing the artificial boundary. The algorithms outlined above comprise the necessary functionality for an incremental Delaunay triangulation

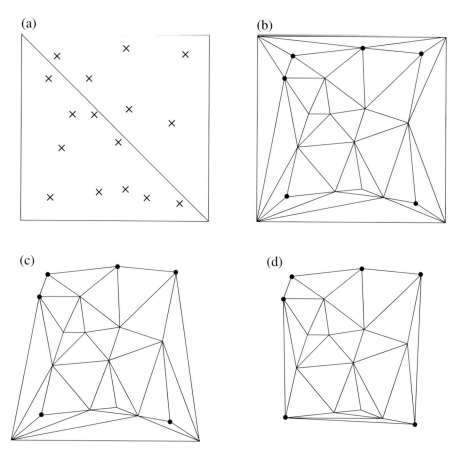

Fig. 9.14. Illustration of the incremental Delaunay triangulation algorithm. (a): two triangles enclosing a point set P; (b): all points in P have been inserted; (c): two nodes of the initial rectangle have been removed; (d): the final Delaunay triangulation. The bullets indicate vertices on the convex hull of P.

algorithm. Starting with an initial triangulation created on the application side, Algorithm 9.3 is called repeatedly for all points in a point set P. The initial triangulation can, for example, be two triangles forming a rectangle enclosing all points of P as in Figure 9.14(a). In our example with the half-edge data structure from Section 9.1, this functionality can be a member function `initTwoEnclosingTriangles` in class `Triangulation`. Figure 9.14(b) shows a Delaunay triangulation $\Delta(P \cup B)$ after all points of P have been inserted, where B denotes the four extra corner points at the boundary.

In most cases one wants the final result to be the Delaunay triangulation $\Delta(P)$ without the four corner nodes of the rectangle. Each node is removed by calling `ttl::removeBoundaryNode(DartType& d)`, where the given dart is one that the actual node is associated with. This function is also implemented generically in TTL (together with the more general function `ttl::removeNode`). It requires a function `removeBoundaryTriangle(d)` in the traits class that removes the boundary triangle t_k associated with a dart $d = (v_i, e_j, t_k)$, without any edge-swapping. A boundary triangle in this context is a triangle with at least one edge at the boundary of the triangulation. In Figure 9.14 we see that `ttl::removeBoundaryNode` may also involve swapping edges near the boundary to maintain the Delaunay property after each deletion. Figure 9.14(c) shows the result after two of the corner nodes have been removed, and in (d) the final result is shown.

Some remarks on implementation. For optimal performance of the incremental algorithm, the points in P may be sorted on their coordinates in lexicographically ascending order as proposed for the divide-and-conquer algorithm in Section 4.8. In C++ this can be done by calling function template `std::sort` in STL if P is represented as a sequence accepted by `std::sort` and if an operator for comparing two elements of P in the lexicographical ordering is provided. (For example, a 2D point (x, y) can be represented by `std::pair<double,double>` and a sequence of 2D points can be represented by `std::vector< std::pair<double, double> >`. In this case, an operator for comparing objects need not be provided since `std::pair` has a "less than" operator that compares two elements lexicographically.) Then the next point to be inserted is likely to be close to the previous one. By returning a dart from Algorithm 9.3 close to the insertion point, this dart can be used as input when the algorithm is called again for the next insertion point. Thus, the average performance of `locateTriangle` in Step 1 would be much faster than $O(N)$, which is the worst case complexity.

In TTL, it was adopted as a general rule that a dart outside a quadrilateral should not be changed as the result of swapping the diagonal of the quadrilateral, while darts inside the quadrilateral could be changed. This was necessary for some of the data structures that were adapted to the generic Delaunay triangulation algorithm. Figure 9.11(b) is in agreement with this rule as the darts $d_{1,1}$ and $d_{1,2}$ are "hidden" outside the quadrilateral where the edge associated with d_1 is swapped before $d_{1,1}$ and $d_{1,2}$ are examined.

Some final remarks on design and performance. Generic algorithms for triangulations can be founded on sound algebraic concepts as defined by G-maps. G-maps provide an algebra with the necessary elementary functions for navigating in the topology of a triangulation at an abstract level independent of a specific underlying data structure. The few and clear basic concepts that G-maps are based on are intuitive and easy to deal with for application programmers. Implementation by means of function templates in C++ separates algorithms from data structures such that applications can adapt arbitrary data structures for triangulations to TTL. Among many useful tools that can be implemented in TTL is incremental Delaunay triangulation as described above. The swapping algorithm for constrained Delaunay triangulation described in Section 6.5 is also implemented generically in TTL.

Algorithms in TTL are compact and easy to read, and thus, easy to maintain and extend with new functionality. Interfaces between TTL and application data structures are clean and narrow with only a few parameters in the argument lists. In fact, many algorithms in TTL pass only one single object in their interfaces: a dart which represents a (node, edge, triangle) triple.

The generic algorithms run without significant loss of efficiency compared to algorithms that work directly on a specific data structure. In particular, the basic topological traversal operators, α_0, α_1 and α_2 of G-maps that dominate the running time of many algorithms, can be implemented as C++ inline functions in a dart class. Compared to abstraction through class inheritance and dynamic binding in C++, the algorithms in TTL are much more efficient.

9.8 Exercises

1. Write dart classes in C++ (or Java) for
   a) the data structure in Section 2.5 and
   b) the data structure in Section 2.6.
   Assume that the fields in the tables representing the data structures are integers that refer to vertex numbers and triangle numbers (indices).
2. Write a pseudo-code based on darts and the α-iterators for a function that finds neighbor vertices of v_i given a dart $d = (v_i, e_j, t_k)$. Write also the corresponding generic C++ code. (You may use an `std::list` to collect vertices.)
3. Write the following function templates in C++. See specification in Section 9.3.
   a) `bool isBoundaryTriangle(const DartType& d)`
   b) `bool isBoundaryNode(const DartType& d)`
   c) `int getDegreeOfNode(const DartType& d)` (the node can also be at the boundary).
4. Analyze the point location algorithm in Section 9.5. Assume that a triangle $t_{i,j,k}$ reached by the algorithm is degenerate and that p lies outside $t_{i,j,k}$ on the extension of the edges, that is, p, p_i, p_j and p_k are collinear. What is returned from the algorithm? Suggest some improvements in this case.
5. Implement a data structure for triangulations, for example the triangle-based data structure in Section 2.5, and adapt it to TTL by implementing the adaptation layer. Check the implementation by running Delaunay triangulation and inserting constrained edges.

6. Write the following class templates in C++ (cf. Section 9.4):
   a) Orbit0_iterator
   b) Orbit1_iterator
   c) Orbit2_iterator
   d) Boundary_iterator
7. Efficient visualization of triangulations in OpenGL using triangle strips and triangle fans, see Section 2.9:
   a) Design an algorithm for preparing a triangulation for visualization in OpenGL using triangle strips (and triangle fans). Use G-maps to traverse the triangulation. Write pseudo codes.
   b) Implement the algorithm by using TTL and a predefined data structure, or use your own data structure (see Exercise 5).

References

1. OpenGL. World Wide Web document at
   `http://www.opengl.org`.
2. OpenSceneGraph. World Wide Web document at
   `www.openscenegraph.org`.
3. Silent Wings Soaring Flight Simulator. World Wide Web document at
   `www.silentwings.no`.
4. Siscat Scattered Data Library. World Wide Web document at
   `www.sintef.no/static/AM/Geom/siscat`.
5. Triangulation Template Library (TTL). World Wide Web document at
   `www.simula.no/ogl/ttl`.
6. E. M. Arkin, M. Held, J. S. B. Mitchell, and S. Skiena. Hamiltonian triangulations for fast rendering. *The Visual Computer*, 12(9):429–444, 1996.
7. M. H. Austern. *Generic Programming and the STL: Using and Extending the C++ Standard Template Library*. Addison Wesley Longman, Reading, MA, 1998.
8. G. Baszenski and L. L. Schumaker. Use of simulated annealing to construct triangular facet surfaces. In P. J. Laurent, A. L. Méhauté, and L. L. Schumaker, editors, *Curves and Surfaces*, pages 27–32. Academic Press, New York, 1991.
9. M. Bern, D. Eppstein, and J. Gilbert. Provably good mesh generation. *Journal of Computer & Systems Sciences*, 48(3):384–409, 1994.
10. Y. Bertrand and J.-F. Dufourd. Algebraic specification of a 3d-modeller based on hypermaps. *Graphical Models and Image Processing*, 56(1):29–60, January 1994.
11. A. Brandt. Algebraic multigrid theory: The symmetric case. *Applied Mathematics and Computation*, 19:23–56, 1986.
12. A. Brandt, S. F. McCormick, and J. Ruge. Algebraic multigrid (AMG) for sparse matrix equations. In D. J. Evans, editor, *Sparsity and its Applications*, pages 257–284. Cambridge University Press, 1984.
13. W. L. Briggs, V. E. Henson, and S. F. McCormick. *A multigrid tutorial: second edition*. Society for Industrial and Applied Mathematics, 2000.
14. L. P. Chew. Guaranteed-quality triangular meshes. Technical Report TR-89-983, Department of Computer Science, Cornell University, Ithaca, NY, 1989.

15. L. P. Chew. Guaranteed-quality mesh generation for curved surfaces. In *Proceedings of the Ninth Annual ACM Symposium on Computational Geometry*, pages 274–280, San Diego, CA, 1993.

16. B. Choi, H. Shin, Y. Yoon, and J. Lee. Triangulation of scattered data in 3D space. *CAD*, 20(5):239–248, June 1988.

17. A. K. Cline and R. J. Renka. A storage-efficient method for construction of a Thiessen triangulation. *Rocky Mountain J. Math.*, 14:119–140, 1984.

18. R. Clough and J. Tocher. Finite element stiffness matrices for analysis of plates in bending. In *Proceedings of the Conference on Matrix Methods in Structural Mechanics*. Wright Patterson A.F.B., Ohio, 1965.

19. M. de Berg, M. van Kreveld, M. Overmars, and O. Schwarzkopf. *Computational Geometry - Algorithms and Applications*. Springer-Verlag, second edition, 2000.

20. L. De Floriani. Data structures for encoding triangulated irregular networks. *Advances in Engineering Software*, 9(3):122–128, 1987.

21. L. De Floriani and E. Puppo. An on-line algorithm for constrained Delaunay triangulation. *Graphical Models and Image Processing*, 54(4):290–300, 1992.

22. B. Delaunay. Sur la sphere vide. *Bulletin of Academy of Sciences of the USSR*, pages 793–800, 1934.

23. N. Dyn, I. Goren, and S. Rippa. Transforming triangulations in polygon domains. *Computer Aided Geometric Design*, 10(6):531–536, December 1993.

24. N. Dyn, D. Levin, and S. Rippa. Algorithms for the construction of data dependent triangulations. In J. Mason and M. Cox, editors, *Algorithms for Approximation II*, pages 185–192. Chapman and Hall, 1990.

25. N. Dyn, D. Levin, and S. Rippa. Data dependent triangulations for piecewice linear interpolation. *IMA Journal of Numerical Analysis*, 10:137–154, 1990.

26. H. Edelsbrunner. *Geometry and Topology for Mesh Generation*. Cambridge University Press, 2001.

27. F. Evans, S. S. Skiena, and A. Varshney. Completing sequential triangulations is hard. Technical report, Dep. of Computer Science, State University of New York, 1996.

28. G. Farin. *Curves and Surfaces for Computer Aided Geometric Design*. Academic Press, Boston, 1988.

29. M. S. Floater. Parametrization and smooth approximation of surface triangulations. *Computer Aided Geometric Design*, 14:231–250, 1997.

30. M. S. Floater. How to approximate scattered data by least squares. Technical Report STF42 A98013, SINTEF, Oslo, 1998.

31. S. Fortune. A sweepline algorithm for voronoi diagrams. *Algorithmica*, 2:153–174, 1987.

32. R. Franke. Scattered data interpolation: Test of some methods. *Math. Comp.*, 38:181–200, 1982.

33. G. H. Golub and C. F. Loan. *Matrix Computations*. John Hopkins University Press, Baltimore and London, third edition, 1996.

34. L. Guibas and J. Stolfi. Primitives for the manipulation of general subdivisions and the computation of Voronoi diagrams. *ACM Transaction on Graphics*, 4(2):74–123, 1985.

35. L. J. Guibas, D. E. Knuth, and M. Sharir. Randomized incremental construction of Delaunay and Voronoi diagrams. *Algorithmica*, 7:381–413, 1992.

36. I. Guskov. Multivariate subdivision schemes and divided differences. Technical report, Princeton University, 1998.

37. I. Guskov, W. Sweldens, and P. Schröder. Multiresolution signal processing for meshes. In *Proceedings of the 26th annual conference on Computer graphics and interactive techniques*, pages 325–334. ACM Press/Addison-Wesley Publishing Co., 1999.

38. Y. Halbwachs, G. Courrioux, X. Renaud, and P. Repusseau. Topological and geometric characterization of fault networks using 3-dimensional generalized maps. *Mathematical Geology*, 28(5):625–656, 1996.

39. Y. Halbwachs and Ø. Hjelle. Generalized maps in geological modeling: Object-oriented design of topological kernels. In H. P. Langtangen, A. M. Bruaset, and E. Quak, editors, *Advances in Software Tools for Scientific Computing*, pages 339–356. Springer-Verlag, December 1999.

40. D. Hansford. The neutral case for the min-max triangulation. *Computer Aided Geometric Design*, 7(5):431–438, 1990.

41. M. Heller. Triangulation algorithms for adaptive terrain modelling. In *Proceedings of the 4th International Symposium on Spatial Data Handling*, pages 163–174, July 1990.

42. M. Hestenes and E. Stiefel. Methods of conjugate gradients for solving linear systems. *J. Res. Nat. Bur. Stand.*, pages 409–436, 1952.

43. Ø. Hjelle and M. Dæhlen. Multilevel least squares approximation of scattered data over binary triangulations. *Computing and Visualization in Science*, 8(2):83–91, 2005.

44. J. Hu, C. Tong, and R. S. Tuminaro. ML 2.0 smoothed aggregation user's guide. Technical Report SAND2001-8028, Sandia National Laboratories, Albuquerque NM, 2000.

45. B. Joe and C. A. Wang. Duality of constrained voronoi diagrams and delaunay triangulations. *Algorithmica*, 9:142–155, 1993.

46. D. Johnson, C. Aragon, L. McGeoch, and C. Schevon. Optimization by simulated annealing: An experimental evaluation; part 1, graph partitioning. *Operations Research*, 37(6):865–893, November-December 1989.

47. L. Kettner. Designing a data structure for polyhedral surfaces. In *The 14th ACM Symp. On Computational Geometry*, pages 146–154, Minneapolis, Minnesota, June 1998.

48. L. Kobbelt, S. Campagna, J. Vorsatz, and H.-P. Seidel. Interactive multiresolution modeling on arbitrary meshes. In *Proceedings of the 25th annual conference on Computer Graphics and Interactive Techniques*, pages 105–114, 1998.

49. C. L. Lawson. Transforming triangulations. *Discrete Mathematics*, 3:365–372, 1972.

50. C. L. Lawson. Software for C^1 surface interpolation. In J. Rice, editor, *Mathematical Software III*, New York, 1977. Academic Press.

51. D. T. Lee and A. K. Lin. Generalized delaunay triangulation for planar graphs. *Discrete & Computational Geometry*, 1:201–217, 1986.

52. D. T. Lee and B. J. Schachter. Two algorithms for constructing a Delaunay triangulation. *International Journal of Computer and Information Sciences*, 9(3):219–242, 1980.

53. P. Lienhardt. Subdivision of n-dimensional spaces and n-dimensional generalized maps. In *5th ACM Symposium on Computational Geometry*, pages 228–236, Saarbrucken, Germany, 1989.

54. P. Lienhardt. Topological models for boundary representation: A survey. Technical report, University of Luis Pasteur, Strasbourg, February 1990.

55. P. Lindstrom, D. Koller, W. Ribarsky, L. F. Hodges, N. Faust, and G. A. Turner. Real-time, continuous level of detail rendering of height fields. In *ACM SIGGRAPH 96*, pages 109–118, August 1996.

56. P. Lindstrom and V. Pascucci. Terrain simplification simplified: A general framework for view-dependent out-of-core visualization. *IEEE Transactions on Visualization and Computer Graphics*, 8(3):239–254, 2002.

57. G. C. M. Behzad and L. Lesniak-Foster. *Graphs and Digraphs*. Wadsworth, Belmond, CA, 1979.

58. J. A. Mchugh. *Algorithmic Graph Theory*. Prentice-Hall Inc., 1990.

59. D. H. McLain. Two dimentional interpolation from random data. *Computer Journal*, 19(2):178–181, 1976. (Errata in 19(4) Nov. 1976, p. 384).

60. S. Meyers. *Effective C++, 50 Specific Ways to Improve Your Programs and Designs*. Addison-Wesley, Reading, MA, 1992.

61. T. Midtbø. *Spatial Modelling by Delaunay Networks of Two and Three Dimentions*. PhD thesis, University of Trondheim, 1993.

62. T. Midtbø. Removing points from a Delaunay triangulation. In *Proceedings From the 6th International Symposium on Spatial Data Handling*, pages 739–750, September 1994.

63. G. L. Miller, S. E. Pav, and N. J. Walkington. When and why Ruppert's algorithm works. In *Twelfth International Meshing Roundtable*, pages 91–102, 2003.

64. A. Mirante and N. Weingarten. The radial sweep algorithm for constructing triangulated irregular networks. *IEEE Computer Graphics and Applications*, 2(3):11–21, 1982.

65. N. Myers. A new and useful template technique: Traits. *C++ Report*, 7(5):32–35, June 1995.

66. R. B. Pajarola. Large scale terrain visualization using the restricted quadtree triangulation. In *Proceedings of the Conference on Visualization '98*, pages 19–26. IEEE Computer Society Press, 1998.

67. M. Perrin, B. Zhu, J.-F. Rainaud, and S. Schneider. Knowledge-driven applications for geological modeling. *Journal of Petroleum Science and Engineering*, 47(1):89–104, 2005.

68. F. P. Preparata and M. I. Shamos. *Computational Geometry, an Introduction*. Springer-Verlag, New York, 1985.

69. W. H. Press, B. P. Flannery, S. A. Teukolsky, and W. T. Vetterling. *Numerical Recipes in C*. Cambridge University Press, 1992.

70. S. Rippa. Adaptive approximation by piecewise linear polynomials on triangulations of subsets of scattered data. *SIAM Journal on Scientific and Statistical Computing*, 18(3):1123–1141, 1992.

71. S. Rippa. Long and thin triangles can be good for linear interpolation. *SIAM Journal on Numerical Analysis*, 29(1):257–270, February 1992.

72. S. Röttger, W. Heidrich, P. Slusallek, and H.-P. Seidel. Real-time generation of continuous levels of detail for height fields. In V. Skala, editor, *Proceedings*

of 1998 International Conference in Central Europe on Computer Graphics and Visualization, pages 315–322, 1998

73. J. Ruppert. A delaunay refinement algorithm for quality 2-dimensional mesh generation. *Journal of Algorithms*, 18(3):548–585, 1995.

74. S. Schneider. *Pilotage Automatique de la Construction de Modèles Géologiques Surfaciques*. PhD thesis, Université Jean Monnet et Ecole Nationale Supérieure des Mines de Saint-Etienne, 2002.

75. L. L. Schumaker. Computing optimal triangulations using simulated annealing. *Computer Aided Geometric Design*, 10:329–345, 1993.

76. M. I. Shamos and D. Hoey. Closest-point problems. In *Proceedings of the 16th Annual IEEE Symposium on the Foundations of Computer Science*, pages 151–162, October 1975.

77. J. R. Shewchuk. Triangle: Engineering a 2d quality mesh generator and delaunay triangulator. In M. C. Lin and D. Manocha, editors, *Applied Computational Geometry: Towards Geometric Engineering*, volume 1148 of *Lecture Notes in Computer Science*, pages 203–222. Springer-Verlag, may 1996. From the First ACM Workshop on Applied Computational Geometry.

78. J. R. Shewchuk. Adaptive precision floating-point arithmetic and fast robust geometric predicates. *Discrete & Computational Geometry*, 18(3):305–363, October 1997.

79. J. R. Shewchuk. *Delaunay Refinement Mesh Generation*. PhD thesis, School of Computer Science, Carnegie Mellon University, Pittsburgh, Pennsylvania, 1997. Available as Technical Report CMU-CS-97-137.

80. J. R. Shewchuk. Delaunay refinement algorithms for triangular mesh generation. *Computational Geometry*, 22:21–74, 2002.

81. R. Sibson. Locally equiangular triangulations. *Computer Journal*, 21(3):243–245, August 1978.

82. B. L. Stephane Contreaux and J.-L. Mallet. A celluar topological model based on generalized maps: The GOCAD approach. In *GOCAD ENSG Conference. 3D Modelling of Natural Objects: A Challenge for the 2000's*, Nancy, June 1998.

83. A. Thiessen and J. Alter. Precipitation averages for large areas. *Monthly Weather Review*, 39:1082–1084, 1911.

84. L. Velho and J. Gomes. Variable resolution $4 - k$ meshes: Concepts and applications. *Computer Graphics Forum*, 19(4):195–212, 2000.

85. M. Von Golitschek and L. L. Schumaker. Data fitting by penalized least squares. In J. C. Mason and M. G. Cox, editors, *Algorithms for Approximation II*, pages 210–227. Chapman & Hall, 1990.

86. D. F. Watson. Computing the n-dimensional Delaunay tessellation with application to Voronoi polytopes. *The Computer Journal*, 24(2):167–172, 1981.

87. K. Weiler. Edge based data structures for solid modeling in curved-surface environments. *IEEE Computer Graphics and Applications*, 5(1):21–40, January 1985.

88. R. J. Wilson. *Introduction to Graph Theory*. Longman, Essex, UK, 1985.

89. M. Woo, J. Neider, T. Davis, and D. Shreiner. *OpenGL Programming Guide. The Official Guide to Learning OpenGL, Version 1.2*. Addison-Wesley Pub Co, third edition, 1999.

Index